Real-life Problems for Introductory General Physics

3rd edition

by Frank Weichman

Real-life Problems for Introductory General Physics, 3rd edition
by Dr. Frank Weichman
ISBN: 1-894380-00-2

FP Hendriks Publishing Ltd.

Canada
4806–53 St.
Stettler, Alberta, Canada T0C 2L2

Phone/Fax: 403-742-6483
Toll Free Phone/Fax: 1-888-374-8787
E-mail: editor@fphendriks.com
Website: www.fphendriks.com

Canadian Cataloguing in Publication Data
Weichman, Frank, 1930–
Real-life problems for introductory general physics

ISBN 1-894380-00-2

1. Physics—Problems, exercises, etc. I. Title.
QC32.W433 2000 530'.076 C00-910297-3

Production Team

Thanks to all those talented people who worked on this project:

Author	Dr. Frank Weichman
Solutions	Kalen Martin, Stephen Steciw, Frank Weichman
Physics Validation	Chris Crawford, Peter Weichman, M. Razavy
Project Director	Faye Boer
Editors	Donna Bennett, Faye Boer
Page Layout and Design	Faye Boer, Leslie Stewart
Illustrations	Dave Fortin, Gabriel Lasarick, Brandi Burns
Cover	Kerry Plumley

Manufacturer
Transcontinental Printing Inc.

PRINTED IN CANADA

Table of Contents

Introduction	**iv**
Chapter One—Mechanics	**1**
Kinematics	1
Forces and Acceleration	5
Circular Motion	9
Work, Energy and Power	14
Momentum	19
Statics	21
Angular Momentum	23
Chapter Two—Simple Harmonic Motion	**29**
Simple Repetitive Motion	30
Energy Considerations in Simple Harmonic Motion	37
Chapter Three—Fluids	**43**
Buoyancy	43
Static Pressure	46
Fluid Flow	48
Viscous Flow	48
Chapter Four—Heat and Thermodynamics	**53**
Temperature and Energy	53
Heat Conductivity	55
Ideal Gas Laws and Kinetic Theory	60
Heat Engines	62
Chapter Five—Wave Motion and Sound	**67**
Frequency, Wavelength, and Speed of Sound	67
Dissipation and Attenuation of Sound	73
Waves of Unspecified Shape	76
Sinusoidal Waves	79
Doppler Effect	85
Standing Waves	89
Beats	92
Chapter Six—Electricity and Magnetism	**95**
Electrostatics	95
Circuits	97
Ohm's Law	99
Ohm's Law and Material Properties	101
Non-ohmic Behavior	104
Magnetism	109
Alternating Currents	113
Chapter Seven—Geometrical Optics	**119**
Rays	119
Light Levels	125
Reflection	128
Refraction	137
Mirrors and Lenses	145
Optical Instruments, Multiple Lens Systems	154
Chapter Eight—Physical Optics	**161**
Interference	161
Diffraction	166
The Speed of Light	168
Polarization	172
Frequency in Electromagnetic Radiation	175
Photons	175
More Optical Instruments	177
Appendix—Numerical Answers	**183**

Real-life Problems for Introductory General Physics

3rd edition

INTRODUCTION

Rationale

This set of problems is designed for use in an introductory level college physics course. The required mathematical knowledge is confined to algebra and trigonometry. Most of the questions are in multiple steps, to lead the students toward the correct approach to problems which otherwise may be intimidating. The assumption also is that students will have access to simple, substitute-a-number-for-a-variable type problems from their textbook for drill purposes.

A strong effort has been made to direct the problems toward illustrating realistic situations to give students a better understanding of the world and the technology around them. Wherever possible physical constants are quoted to the full accuracy of current measurements or knowledge. For some of the problems this is important, for others it is meant as a reminder that science is not restricted to a precision of three significant figures.

The intent is for students to develop skills by using physics concepts and principles to solve real-life problems. These concepts and principles can be found in any introductory physics textbook and/or classroom notes.

Format

Each chapter begins with a brief introduction. The purpose of the introduction is to provide background information or to provide a context for the topics or problems to follow. Introductory passages also appear throughout the chapter when further background information is required for the problems immediately following. The introductory passages are not intended as a substitute for a textbook or for classroom instruction. They may, however, help students find the sections of the text or classroom notes which are needed to solve the problems.

In a few cases equations appear in the introductory paragraphs or in the problems themselves which do not come from any standard physics text. The purpose of the problems that follow is to extend the range of the students' experience with problem solving. Memorizing those equations is not required or even suggested.

Finally there are some problems in which repetitive calculations based on the same equation are required. The reason for asking such questions is that they illustrate interesting trends. With computers so widely available now, it is hoped that these questions will encourage the student to write a simple program to do the repetitive calculations. However, a simple calculator that has trigonometric and logarithmic functions is all that is really required.

Illustrations

Many of the problems have small illustrations to assist students in visualizing the situation described in the problem. These illustrations are referred to as figures and identified with the chapter number and then a number which identifies the order the illustration appears in the chapter. For example, Figure 6.2 is the second illustration in chapter 6.

Written Response Icon

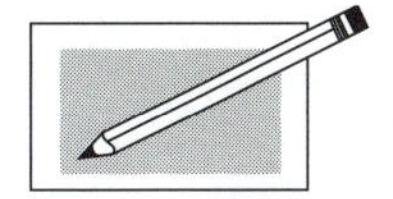

This icon appears beside problems where a fairly lengthy written response is required in addition to, or instead of, a calculated response.

Numerical Answers

Numerical answers are provided for all numerical problems to give students confidence that they are on the right track in solving the problems. Remember though that understanding the logic and the method of approach are always more important in physics than obtaining the correct numerical answer.

NOTES

I MECHANICS

A GENERAL COMMENT

The use of kilometers per hour (km/h) instead of the SI units of meters per second (m/s) in many problems is deliberate. Students and the general public are familiar with numbers associated with speed and velocity almost exclusively from their own experiences in motorized vehicles. Therefore, they have developed a sense of speeds in km/h as being fast or slow. It is hoped that the conversion factor of 3.6 between km/h and m/s will also soon become intuitive. Similarly we have, on occasion, used weight in pounds (lb) instead of mass in kilograms (kg). Again the reason is that, in spite of our federal laws, we still worry about our weight in pounds and buy our groceries in the same units. For some of the problems in which pounds are used no conversion is necessary because only ratios are needed.

1.1 KINEMATICS

In spite of the push to have SI units introduced for international uniformity and simplicity, old units persist in the working world and in the scientific world. In most cases familiar units persist where they make comparisons easy. After a party you might compare the glasses of wine you drank, or the number of bottles of beer, not the alcohol consumption as measured in m^3. Pressures should be measured in N/m^2, but the atmosphere as the pressure unit is often preferred. Astronomers quote the distance to stars in light years instead of meters. Navigators use knots instead of m/s. Do you give your age in seconds or in years?

1.1.1 On shipboard and in aircraft speeds are preferably measured in knots instead of m/s. The for example, the speed of a merchant ship is 15 to 20 knots, while the speed of a jet aircraft is close to 500 knots.

The definition of knot is one nautical mile per hour. The nautical mile in turn has the same origin as the meter, that is, it is a well-defined fraction of the circumference of the Earth. The original definition of the meter was one part in 40.00×10^6 of the circumference of the Earth. The length of the nautical mile is obtained from the circumference of the Earth by dividing the 360° of the circle into minutes of arc. Each minute of arc is then one nautical mile. The nautical mile is therefore the circumference of the Earth divided by 360 x 60.

a. Determine the length of one nautical mile in meters.

b. Convert one knot into meters per second (m/s).

c. A ship travels at 18 knots. Calculate its speed in kilometers per hour (km/h). Repeat for a jet aircraft moving at 490 knots.

d. In the days of sailing ships, the speed of the ship was determined by throwing overboard a float (the log) attached to a long line. The line had equally spaced markers (knots) on it. The sailor counted the number of markers that slipped through his hands in a given time interval once the log was at rest with respect to the water. The number of markers counted was the speed of the ship as measured in knots. Given that the time interval for the measurement was 28 seconds,* what was the distance between the markers on the line? In those days the foot was the standard unit of distance. Show that 47.25 feet was the appropriate distance between the markers.

1.1.2 The light year (ly) and the astronomical unit (AU) are the preferred units in astronomy. For distances inside the solar system the astronomical unit at $1.495\,978\,70 \times 10^{11}$ m is preferred because it is the mean radius of the Earth around the Sun. For interstellar distances the preferred unit is the light year . The closest stars are 4 ly from the solar system. The light year is defined as the distance light travels in one year in a vacuum. The speed of light in a vacuum is 299 792 458 m/s.

a. Pluto is most of the time the outermost planet in our solar system, with a mean orbital radius of 39.44 AU. Convert this distance to meters and light years.

b. The brightest star visible from the Earth is Sirius. Its distance from the solar system was determined to be 8.7 ly. Convert this distance to meters and astronomical units.

1.1.3 Some conversions are strictly for laughs. Physicists have a strange sense of humor. Here are two to try your hand at.

a. How many minutes are there in a microcentury (micro = one millionth)?

b. A furlong is 1/8 of a mile. A fortnight is two weeks. What is the speed of light in furlongs/fortnight?

1.1.4 The specifications for aircraft performance include the concept of glide ratio. It is the ratio between the horizontal distance the aircraft covers without power for the vertical distance it covers at the same time while in controllable flight at sufficient air speed. The glide ratio is small for

*Encyclopedia Brittanica, 1968.

commercial aircraft and much larger for sail planes—gliders. The space shuttle glides to a landing after space flights and even a Boeing 747 was brought to a safe landing this way after running out of fuel over Gimli, Manitoba.

A glider is released from its tow at an altitude of 875 m in still air. In flight it will have a forward speed of 53.6 km/h while losing altitude at 0.470 m/s.

a. What will be the glide ratio of the aircraft under the above conditions?

b. What will be the range of that glider under still-air conditions?

c. The local club of glider pilots holds a contest on a day when the wind is at a steady 6.10 m/s from the north, independent of altitude. The gliders start from the same height of 875 m and, after a straight-line flight, land at the same distance from the release point as they would have reached after a straight-line flight, without wind. In what direction(s) must the pilot choose to fly? Where will the aircraft land with respect to the release point?

1.1.5 Passing a car on a two-lane road requires increasing caution as speeds increase. The primary requirement is that there be enough space on the road between oncoming vehicles. Experience shows that this space requirement increases with increasing speed of traffic.

For simplicity's sake let the length of all cars be 4.1 m. Suppose further that a reasonably courteous maneuver starts with the passing driver at least three car lengths behind the slower car and that the driver pulls in with the back of his car at least three car lengths ahead of the slower car. Also assume the passing car moves 11 km/h faster than the slower car.

a. Begin with the simplest case. The slower car is standing still; the passing car moves at 11 km/h. Calculate the distance required for the one car to pass the other with appropriate spacing. How many seconds will the maneuver take?

b. Prove that the time it takes for one car to pass the other depends only on their relative speeds.

c. The slower car moves at 100 km/h while the passing car moves at 111 km/h. The same spacing between the cars at the beginning and at the end of the maneuver is assumed. Calculate the distance along the highway that is required to be free of oncoming traffic at this speed.

1.1.6 Getting from Point A to Point B involves decision making on the part of a driver and on the part of transportation engineers who plan the split between neighborhood roads, arterial roads, and freeways. Local residents want their neighborhoods free of traffic but they also want access, in the shortest possible time, to any place in the city they wish to go. The accepted compromise is to slow speed deliberately in neighborhoods using many stop signs and streets with continually changing directions. Arterial roads are at the edges of residential communities. These roads are wider and traffic lights are used to ensure a reasonable flow of traffic at higher speeds. For transportation over longer distances, but still within the context of the large city, limited access highways (freeways) are built where speeds are similar to those on intercity highways. On freeways traffic can enter, leave, or cross over the traffic flow without stopping.

Figure 1.1 shows the three types of roads through Metro City. A high speed, circular by-pass freeway, one of the arterial roads, and one direct road through residential communities. The radius of the freeway is 8.7 km as measured from the center of town.

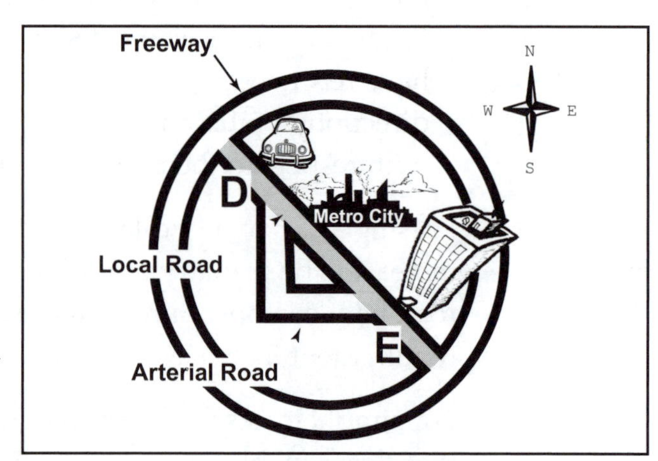

Figure 1.1

Digby has three options to get from his residence to Elvira's apartment. He could drive straight through the center of town for a total distance of 15.8 km. However, his average speed would be 25 km/h because he must obey stop signs and traffic lights as well as avoid pedestrians. He could also drive 0.8 km to the nearest arterial road, make one more turn to get to the intersection of the local road, and then drive an additional 0.8 km to reach Elvira. The third option is to start off in the opposite direction, again for 0.8 km, go to the freeway, follow the freeway halfway around town, and then drive 0.8 km on the local road to Elvira's. The average speed on the arterial roads is 50 km/h; the average speed on the freeway is 90 km/h.

a. Calculate the time it takes to drive each of the three routes from Digby's to Elvira's.

b. Calculate the average velocity for each of the three proposed routes.

1.1.7 At-grade Light Rail Transit (LRT) lines are being built in cities where underground systems are too expensive for the expected number of passengers. Because they have their own right-of-way, they operate more independently of traffic conditions than do buses. Because they usually operate at street level and without fencing, their speed is kept low. For a typical LRT system, the stations are 800 m apart and the maximum operating speed is 70.0 km/h. For the purposes of this problem assume, as a good approximation to actual fact, that the trains accelerate and brake at 1.00 m/s^2. (For emergencies the brakes are capable of causing negative accelerations of 2.5 m/s^2.)

For the following assume a clear right-of-way where the trains are permitted to go at their maximum rated speed.

a. How long does it take for the train to reach its maximum operating speed?

b. How far does the train travel in this time?

c. How long does it take for the train to get from one station to the next?

d. Station stops are scheduled to take 20.0 s each. Under the above conditions, determine the speed of the trains averaged over a number of stations.

e. Because of safety considerations in congested areas, the maximum speed of the trains is kept to 60.0 km/h over a stretch of tracks in a residential area. Under these conditions, determine the speed of the trains averaged over a number of stations.

1.2 FORCES AND ACCELERATION

There are a class of situations where neither precise acceleration nor precise force is important, but where an average value suffices to predict devastation. Some examples are the use of seat belts and air bags in vehicles and the requirement to have foam-filled mats at the landing areas of pole vault and high jump events. Any change in velocity is due to a force, but an estimate of the acceleration during that change in velocity is enough to determine how dangerous the situation is. As a rough guide to the damaging effects of acceleration, keep in mind that your feet carry your full weight $m\mathbf{g}$ while you stand still. If you stand on a platform accelerating upward at 9.8 m/s^2, then your feet must support a force twice as great, $2m\mathbf{g}$. In other words, your feet will carry twice the normal weight. Potential astronauts and fighter pilots in training are exposed to extreme accelerations and apparent weightlessness. Accelerations over 5 times **g** can bring on loss of consciousness and accelerations of 10**g**, even on well-padded couches, seriously distort the body. Without the couch and without careful placing of the body, bones can break.

1.2.1 Parachute jumpers and skydivers usually land on solid ground. They are trained to collapse their bodies appropriately on landing. Estimate an acceptable landing speed by considering the distance over which the body can collapse in a fall. From what height would one attain this speed without a parachute?

1.2.2 The maximum height cleared by a pole-vaulter is now close to 6 m. As in the high jump, the emphasis is on clearing the bar without much thought of a perfect landing on the other side. The athlete usually lands horizontally on a high pile of foam-filled mats rather than vertically on solid ground.

a. If the athlete were to land at the same level as the take-off, then at what speed would he land after a pole vault of 5.80 m? Imagine that after this Olympic achievement the athlete landed on his feet and broke his fall by "slowly" collapsing on bent knees over a distance of 77 cm. What would be the average acceleration of the body of the athlete? How many **g**s would the athlete experience?

b. Estimate how many multiples of **g** the athlete would experience if he fell sideways on a 5.0 cm thick mat. Would that be damaging to the athlete's body? Why?

c. In practice, there are layers of foam-filled mats about a meter thick to break the athlete's fall. Answer the same question as in *part b* for these more realistic conditions. How many **g**s would the athlete experience?

1.2.3 Freight trains on the prairies can have 100 cars or more. These long trains usually have more than one locomotive and can be more than 1.5 km long. A single freight car can carry a load of 100 tonnes (1 tonne = 1000 kg). Crossing mountainous terrain, such as the Rockies between Banff and Revelstoke, additional locomotives are often added in the middle of the train. The crew in the front locomotive controls all other engine units by radio signals.

The freight cars are coupled and the tension in the couplings between the cars varies with the location of the locomotives. Simplifications are required to see a trend. First assume that each locomotive pulls the train with the same force **F** and that each freight car resists motion through friction with a force **f**. The train under consideration has two locomotives and 100 cars. It moves at a constant speed on level ground.

a. Determine the tensions in the coupling pulling the last car and the coupling pulling the first car. What is the tension in the coupling between the two locomotives hooked together at the front? What is the frictional force required between the wheels and rails of each of the locomotives?

b. One of the locomotives is shifted to the center of the train with 50 cars ahead of it and 50 cars behind it. Under these conditions, determine the tension in the coupling pulling the first car and in the coupling pulling the last car. What is the tension in the coupling just ahead of the second locomotive? What is the frictional force required between the wheels and rails of each of the locomotives?

c. An 80-car coal train has entered the mountains and is crawling up an incline of 0.72° with 4 locomotives providing the power (up to 4 000 hp per unit). One pair of locomotives is at the front; a second pair is at the center of the train. Each loaded freight car has a mass of 79 tonnes. The locomotives have a mass of 12 000 kg each. All friction except that required for traction at the wheels of the locomotives is to be ignored. Under these conditions, determine the tension in the coupling at the front of the first car and the tension in the coupling pulling the last car. What is the tension in the coupling just ahead of the second set of locomotives? What is the frictional force required between the wheels and rails of each of the locomotives? Give an estimate of the coefficient of friction required between these wheels and rails.

1.2.4 Ocean tides, as you may know, are due to the gravitational influence of the Moon on the Earth. In 1974 a book called *The Jupiter Effect* was published. The author of the book predicted many disasters in the California earthquake zone because Jupiter was going to be unusually close to Earth in the following year. Simple calculations can be done based on astronomical data and Newton's Law of Gravitation to find the gravitational effects of the Sun, the Moon, and Jupiter on the Earth.

Following is a list of some astronomical data:

Object	**Mass**	**Radius**	**Mean Orbital Radius**
Sun	1.99×10^{30} kg	696 000 km	N/A
Jupiter	1.90×10^{27} kg	143 200 km	778.3×10^{6} km
Earth	5.98×10^{24} kg	6378 km	149.6×10^{6} km
Moon	7.35×10^{22} kg	1738 km	384.4×10^{3} km

Depending on the precision required for physics problems, the value of the gravitational acceleration at the surface of the Earth is usually taken as 9.8 m/s^2, however 10 m/s^2 can be used for rough estimates. Precision determinations are also available for geophysical purposes. For example, as part of a project of the Canadian Gravity Network, the value of **g** on a pier in the basement of the Physics Building of the University of Alberta was determined to be 9.8115310 m/s^2. The currently accepted value of the universal gravitational constant G is 6.67259×10^{11} $m^3/kg\ s^2$.

a. Based on the orbital information given, calculate the contribution of the gravitational attractions of the Sun, the Moon, and Jupiter at a point at the surface of the Earth facing that celestial object. Assume Jupiter is at its closest approach to the Earth, i.e. assume Jupiter, Earth, and Sun are in a straight line and in that order.

b. What fraction do the results obtained in *part a* represent in comparison with the Earth's gravitational acceleration? Would that fraction be measurable with the techniques used for the Canadian Gravitational Network?

c. The ocean tides are due to the difference in the gravitational attraction of the moon at opposite sides of the Earth. What is the percentage difference between the attraction of the moon on the Earth between the side of the Earth facing the moon and the side of the Earth facing away from the moon?

1.2.5 There is a parlor trick in which the performer jerks a tablecloth off a table without disturbing the neat place settings on the tablecloth. The details of how to do the trick and what to expect can be calculated.

The starting point is the physical arrangement. A tablecloth covers d = 1.20 m of a smooth table with the rest of the tablecloth hanging loosely to the floor. On the tablecloth but very close to the edge of the table are plates, cutlery, and cups. The coefficient of friction μ between table settings and tablecloth is 0.0640. The performer takes hold of the hanging end of the tablecloth and pulls so hard that the entire tablecloth moves at a constant speed v = 14.0 m/s. With the numbers as stated, it can be shown that the tablecloth gets pulled from under the place setting with hardly a movement of the place setting. The rest of the problem will justify the statement.

a. Calculate the maximum acceleration of the place setting as the cloth underneath moves at the constant speed of 14.0 m/s.

b. During the time of acceleration, the place setting moves a small distance x toward the edge of the table and the distance the cloth travels while in contact with the place setting is x + 1.20 m. Write an expression for and calculate the numerical value for the time of contact between tablecloth and place setting as the tablecloth is moved.

c. Calculate the distance x the place setting moves while the cloth is being pulled.

d. CHALLENGE: Show algebraically that $x = \mu g d^2/2v^2$ for sufficiently small values of x.

1.3 CIRCULAR MOTION

1.3.1 Back in the days of manned lunar exploration the command module with one astronaut on board orbited the moon at an altitude of 50 km above the lunar surface while the other astronauts landed and explored.

The mass of the Moon is 7.35×10^{22} kg and the radius of the Moon is 1.738×10^6 m. The mass of the Earth is 5.97×10^{24} kg and the radius of the Earth is 6.378×10^6 m.

a. Calculate the orbital period of the command module around the moon.

b. The radius of the Moon is smaller than the radius of the Earth by a factor of more than 3. Why should it take less time for a satellite or the space shuttle to make a circuit 50 km above the surface of the Earth than the command module to make a circuit around the Moon?

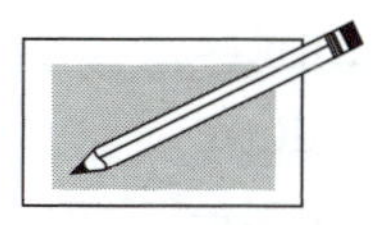

c. The word "weight" has a specific scientific meaning. Explain precisely whether the astronaut in the command module was weightless, massless, or apparently weightless.

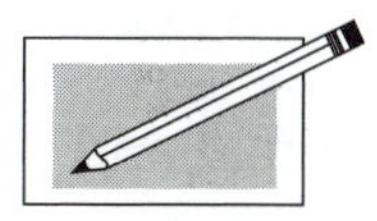

d. How would the exploration of the moon be affected by the phases of the moon, for example, if on the days of the exploration the moon had been in a crescent phase?

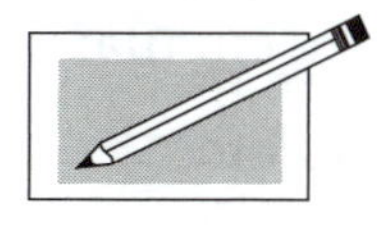

e. In a similar vein, astronauts of an "intellectually challenged" region of the world decide to go on a expedition to do a close-up study the sun. When warned that it gets very hot close to the sun, these would-be explorers explain that their trajectory will take care of that problem. Their closest approach to the sun will take place at night. Comment on the reasoning given.

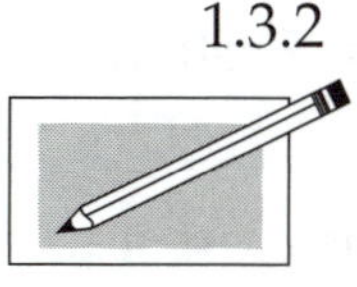

1.3.2 In a James Bond movie Agent 007 has the climaxing fight on an Earth-orbiting space station. The villains are eventually tossed out of the station through an open hatch and are seen to fall rapidly toward Earth.

a. Astronauts use a safety line while working on the outside of a space capsule or the shuttle. Suppose the line comes loose while the astronaut while the astronaut is 5.1 m away from the orbiting laboratory and everyone else is asleep. What must he do to get back? (He only has a few tools he can safely dispose of.) Would he fall rapidly back to Earth as the movie implied? Explain carefully.

Figure 1.2

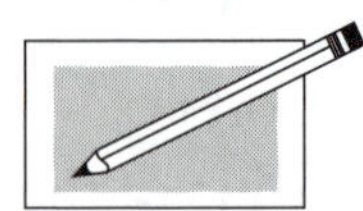

b. US space flights are launched from Florida over the Atlantic Ocean or over the Pacific from California to limit the fall of debris over settled areas in case of failure. Why is the Florida launch preferable to a launch from California? The European nations use a launch site in French Guyana. Why is that location even better than the North American sites?

1.3.3 Two critical unknowns in the early days of astronomy were the mean distance from the Earth to the Sun, called the astronomical unit (AU), and the universal gravitational constant G. The accepted values for these two are 1 AU = $1.495\,978\,70 \times 10^{11}$ m and $G = 6.672\,59 \times 10^{-11}$ m^3/kg s^2. Combine these two with the orbital periods and the basic numbers associated with the Solar System can be determined. This problem guides you through to one of these determinations, the mass of the planet Saturn.

A few words of advice: The required calculations can be simplified and kept to a minimum by doing algebra first and making use of proportions. Assume all orbits of planets and their satellites are circular.

a. Given that the period of Saturn (the time required for a complete circuit around the Sun) is 29.456 yr, calculate the orbital radius of Saturn. Use the known orbital period of the Earth.

b. Earth, Saturn, and most of the other planets are close to orbiting in the same plane around the Sun. They also move in the same direction. Calculate the distance at closest possible approach between Earth and Saturn.

c. The satellites of Saturn can be best observed when Saturn is closest to the Earth. Titan, the largest of the moons, makes a circuit of Saturn in 15.945 days. The maximum angular separation (center-to-center) between Titan and Saturn as seen from Earth is 5.46×10^{-2} degrees. Calculate the orbital radius of Titan and the mass of Saturn.

1.3.4 The four largest moons of Jupiter were discovered by Galileo in 1610. The largest of these moons is Ganymede. It has an orbital period of 7.155 days and a mean orbital radius of 1.070×10^{9} m. The smallest known moon of Jupiter is named Thirteenth. It is barely 15 km across and has an orbital period of 210.6 days.

a. Calculate the mass of Jupiter.

b. Calculate the mean orbital radius of Thirteenth.

c. If an astronaut landed on Thirteenth, then that astronaut would feel weightless, but not so on Ganymede. What is the weight of the astronaut on Thirteenth and in what direction?

1.3.5 In Problem 1.2.4 you were asked to calculate the contribution of the gravitational attraction of the Sun at the location of the surface of the Earth. The magnitude of that contribution is 5.93×10^{-3} m/s^2. It would appear that this contribution should influence measurements of **g** at the surface of the Earth to a noticeable extent, particularly as a 1.2 cm/s^2 day-to-night variation. In fact, the difference barely exists. Just as the astronaut is apparently weightless while orbiting the Earth, all objects on Earth are apparently weightless with respect to the Sun because they are orbiting the Sun.

a. Given that the mean radius of the orbit of the Earth around the Sun is 1.496×10^{11} m and that the centripetal acceleration of the Earth in that orbit is 5.93×10^{-3} m/s^2, calculate the orbital speed of the Earth.

b. Calculate the circumference of the Earth's orbit, assuming a circular path. Using that result calculate the time in seconds and the time in days to complete one orbit.

c. If the daily rotations of the Earth around its own axis are ignored, then the Earth acts as a large object moving around the sun at constant angular velocity with a centripetal acceleration varying slightly from place to place depending on the precise distance from the Sun—the center of the circle. The maximum variation in the distance from the sun is twice the radius of the Earth. Calculate the maximum variation in the gravitational acceleration at the surface of the Earth due to the Sun.

NOTE: The average value of the gravitational acceleration is not noticeable because of the Earth's orbital motion. The variation you have just calculated, which is a local difference from the average, contributes noticeably to the ocean tides.

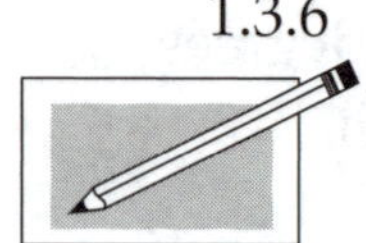

1.3.6 Barbarella is in a large spaceship in interstellar space where gravity is negligible. The spaceship itself is moving along at constant velocity. At an unfortunate moment Barbarella finds herself floating at rest in the spaceship in her space suit. There is no air in the spaceship at that moment and she is well away from all walls. The spaceship suddenly rotates to provide artificial gravity. How, where, and when will this artificial gravity make its presence felt by her? Be as detailed as you can.

1.3.7 Surveyors use lasers and reflectors to measure distances using accurate timing and a precise knowledge of the speed of light. A more sophisticated system of time measurement and the speed of light is the GPS (Global Positioning System).

In this system a hand-held receiver picks up timing signals from a group of high orbiting satellites, does some calculations, and, within seconds, determines the location of the receiver.

The basic principle is simple. Start with a known location A of a satellite and a timed distance R_A to the unknown. This places the unknown at the surface of a sphere specified by the center A and the radius R_A. Add the information that the unknown is at the same time also at a distance R_B from a second satellite, point B. The combined data restricts the location of the unknown to a position where the two spheres A and B intersect. Two intersecting spheres define a circle. Given three known satellite positions A. B, and C and the distances R_A, R_B, and R_C from each of these positions to the unknown gives three intersecting spheres, defining a unique point in space. A fourth position and distance can be used to make corrections for measuring errors.

FOR YOUR INFORMATION: In the GPS application, the hand-held detector can accurately determine the time it takes for a radio signal to travel from each of four orbiting satellites to the detector. Given this travel time, the known speed of light, and the known positions of the four satellites at any time, the position of the detector with respect to the satellites can be determined. A state of the art receiver determines longitude, latitude, and altitude to within 20 cm.

The satellites used for the GPS are highly sophisticated devices. The first requirement is that they be in precisely predictable orbits such that, at the receiver end, their positions are known to a high degree of accuracy at all times. The high circular orbits, about 20 000 km above the surface of the earth, assure the absence of atmospheric friction and therefore, long-term orbital stability. The second requirement is that the satellites identify themselves so the receiver knows which of the 24 satellites it is dealing with. The third requirement is that the satellites send out precise timing information. For this purpose each satellite is equipped with an atomic (Cesium or Rubidium) clock with a stability of 2 parts in 10^{13} or better. The clocks in turn are regularly checked against laboratory standards on earth. The fourth requirement is that the speed of light (more correctly, the speed of the radio waves) is precisely known. The speed of light varies a small but significant amount due to variations in the density of ions in the upper atmosphere. Since the variation changes in a predictable way with broadcast frequencies, each satellite broadcasts its signals at both 1575.42 MHz and 1227.6 MHz. This enables the system to determine the actual speed of light at the time and place of the measurement.

The US Defense Department actually deliberately introduces small, secret variations in the timing signals to prevent foreign missiles from accurately homing in on targets in the US. The US armed forces have the codes to circumvent the error signals. For the commercial and amateur user there are also means available to bypass the deliberate broadcast errors.

a. The closest any of the satellites can be to a receiver on Earth is about 20 000 km. The speed of light is approximately 3×10^8 m/s. Estimate the time it takes for a radio signal to go from the satellite to the receiver. As stated in the introduction, the receiver can be located by the timing signals to within 20 cm. Determine to what parts per million the timing must be reliable in order to achieve the 20 cm location goal. In parts per million, how reliable are the atomic clocks? How well established is the speed of light in parts per million?

b. The GPS satellites are all in circular orbits which have accurately timed periods of 12 hours while the earth rotates under the orbiting satellite. Calculate the required orbital radius of such a satellite.

c. Show that it takes a minimum of 24 hours for any of the satellites to pass over the same spot on Earth again.

d. The satellites are neither in equatorial nor are they in polar orbits. They make an angle of 63° with respect to the plane which contains the Earth's equator. Explain how far north or south of the equator an observer can be for one of these satellites to pass straight overhead.

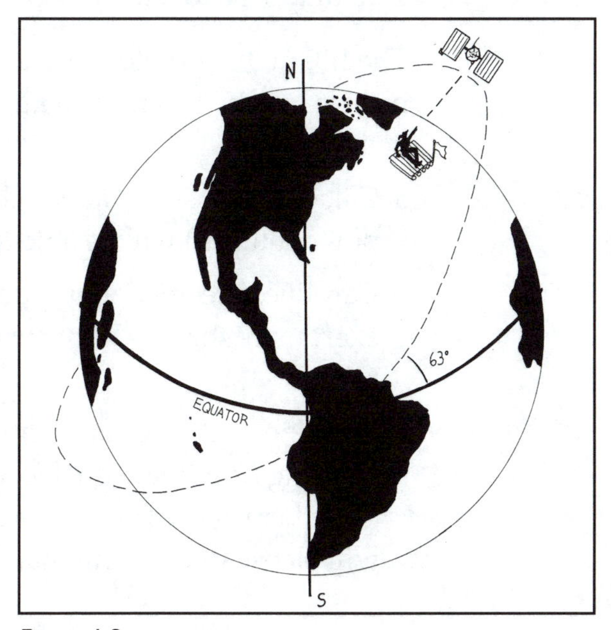

Figure 1.3

1.4 WORK, ENERGY, AND POWER

1.4.1 The 80-car coal train of Problem 1.2.3 is proceeding up the mountain at a grade of 0.72° with 4 locomotives providing power at up to 4 000 hp per unit. One pair of locomotives is at the front; a second pair is at the center of the train. Each loaded freight car has a mass of 79 tonnes (1 tonne = 1 000 kg). The locomotives have a mass of 12 000 kg each. All friction except that required for traction at the wheels of the locomotives is to be ignored.

a. Calculate the maximum speed of the train on the given incline.

b. At an unusually steep section of the railroad right-of-way the grade is 1.33°. Calculate the maximum speed of the train on this section and the coefficient of friction required under the wheels of the locomotives.

1.4.2 A cyclist is about to ride up a long hill. Years of experience have shown that extra speed on the flats before the hill makes getting to the top easier. Experience in riding up hills also shows that the effort required to continue up the hill increases as the speed of the bicycle decreases. This problem analyzes the physics behind the cyclist's observations based on biological data. Specifically, measurements on body metabolism of highly trained marathon runners have shown that such athletes burn fuel in the body at a steady rate of 1.5 kW, but only about 20% of the food energy is converted to the mechanical motion of a given sport. For short bursts of effort such as a jump or a sprint, the human body can generate close to one horsepower in mechanical motion.

The hill in this problem has a slope of 3.0°. The mass of the cyclist plus bicycle is to be taken as 68 kg. All losses associated with friction are to be ignored.

a. Calculate the speed at which the trained cyclist (300 W mechanical power output) will be able to cycle up the hill.

b. Caroline, a recreational cyclist on a family outing, would not wish to exert more than 60 W of steady power. How fast will she go up the hill?

c. Caroline next takes a run at the hill, starting up at 20 km/h and coasting from there on. How far, as measured along the hill, will she go as her speed drops from 20 km/h to 19 km/h? How much time will elapse in covering that distance?

d. Higher up the hill Caroline's speed will have slowed to 5.0 km/h. How far, as measured along the hill, will she go as her speed drops from 5.0 km/h to 4.0 km/h? How much time will elapse in covering that distance?

e. Once the speed of the bike has dropped to walking speed most cyclists dismount and push the bike up the hill. The hill in the above problem is only 5.50 m high. Caroline would like to ride up the hill without losing all speed. She therefore takes a run at the hill to have a greater amount of kinetic energy at the bottom of the hill. She starts up the hill at 36.0 km/h and by additional pedaling she reaches the top of the hill with a speed of 3.0 km/h, just barely a walking pace. How much energy does she have to supply in order to climb the hill? What average power does this represent?

f. Caroline starts up the hill a third time at the more leisurely pace of 12.0 km/h. Again she pedals to keep her speed from dropping to less than 3.0 km/h. How much energy does she have to supply by pedaling to get up the hill? What is the average power she must supply?

1.4.3 Problem 1.4.2 was based on the unrealistic assumption of a complete lack of friction. Friction is an important aspect of all forms of transportation. Measurements of the power required to operate a typical bicycle have established an approximate expression of the frictional force on a bicycle and its rider in still air as $f = Am + Bv^2$. In this expression m is the mass of the bicycle plus rider, v is the speed, and A and B are empirical constants to be taken as 0.050 and 0.20 respectively. The term Am is the contribution of friction which is independent of speed, while the term Bv^2 takes air resistance into account—air resistance increases rapidly with increasing speed. The constant A will be larger when the air pressure of the tires is low, when the road is rough, and when the oil and grease on the chain and on the bearings of the wheels are dirty. The constant B can be decreased with better streamlining and tight-fitting clothing. The power required to ride at a given speed on level ground will be $P = fv = Amv + Bv^3$.

a. On the basis of the above information, calculate the power output required to ride a bicycle at 18 km/h and at 42 km/h. Cyclist and bicycle together have a mass of 85 kg.

b. The maximum speed of a cyclist coasting down an incline can be calculated from the component of the gravitational force acting along the incline and equating that to the frictional forces on the bike and rider going downhill. Calculate the maximum coasting speeds expected from two cyclists of 53 kg and 85 kg respectively (mass of bicycle included) on two hills. One hill has a 3.0° incline and the other hill has a 6.0° incline.

c. To ride uphill takes power to overcome both gravity and friction. It takes too much algebra to calculate the speed attained for a given power input. It is simpler to determine the power required to go at a given speed. Calculate the power requirements for the 85 kg cyclist (bike included) to go up the 3.0° and 6.0° inclines at steady speeds of 3.5 m/s and 7.0 m/s respectively.

1.4.4 The power of engines in cars and trucks is part of the usual sales pitch from the car salesman. For cars the power range is typically between 80 hp and 200 hp at 750 W per metric hp. The power of truck engines is barely a factor of two higher. The power that is actually available to move the cars and trucks is only about 1/7 of the rating of the engine. The power of the engine is also highly dependent on the speed at which the motor turns, which is the reason there are manual or automatic transmissions in all cars. For the purposes of this problem some simplifying assumptions are used to ease the calculations.

The car and driver in this question have a total mass of 720 kg. The quoted power is 98 hp of which only 1/7 is actually available for driving. It is to be assumed that the available power does not depend on the speed of the car. Air resistance is to be ignored.

a. The full available output power of the car is applied to accelerate it. What acceleration is the car capable of at 20 km/h, 60 km/h, and 120 km/h?

b. How long will it take to speed up by 1.0 km/h at 20 km/h, 60 km/h, and 120 km/h? What distances does the car travel in these time periods?

c. Add three adult passengers at 75 kg each, a trunk full of suitcases with a total mass of 120 kg, and a roof rack with ski equipment with an additional mass of 30 kg. Repeat the calculations of *part a* for the car with this extra load.

d. Before the rise of environmental consciousness and energy conservation, cars were heavier and larger. Big engines were also popular. Consider the following example from those good, old pre-metric days. A "muscle car" had a weight of 3200 lb and was able to accelerate from 0 to 60 mph in 9.7 s. Calculate the power required to achieve that acceleration. Keeping in mind the factor of 7 inefficiency, estimate the advertised horsepower of the car. Use the conversions 1.0 kg = 2.2 lb and 1.0 mph = 1.6 km/h.

1.4.5 A ski lift takes skiers to a height 412 m above their starting point over an average slope of 33.5°. The lift consists of a series of equally-spaced benches, for two skiers each, hung on a moving cable in a long loop. The mass of each bench is 12.3 kg and the average mass of each skier including gear is 86.4 kg.

a. How much energy must the lift motor provide for each chair full of skiers?

b. The lift is said to have a capacity of 750 skiers per hour. Assuming perfect efficiency, calculate the minimum power rating for the lift motor.

c. At full power and at maximum capacity, how fast are the chairs moving if they are 11.0 m apart?

d. The stationary skiers are scooped up by the moving chair at the bottom of the lift. They slide off the chairs at the top of the lift at the speed of the chair. How much kinetic energy must the lift provide for each pair of skiers? How does that compare with the potential energy the lift must provide?

e. How long does it take for the individual skier to get to the top of the lift?

f. Suppose the average skier takes an average of 8.8 minutes, including a spill or two, to ski back to the base of the lift. How many skiers can the lift accommodate with zero waiting time? If there were twice that many skiers, how long would skiers have to wait in line to get on the lift again?

1.4.6 Energy consciousness is becoming increasingly important and is encouraged by governments at all levels. One reason is economics. Energy efficiency lowers the cost of operating devices; it can also save money for utilities by postponing the building of new electric power plants. A second reason is that, with few exceptions, the use of energy creates pollution. Cars, refrigerators, light bulbs are but some of the devices that have energy efficiency ratings. Efficiency ratings in turn depend on how the device is used. A super-insulated house can be a great energy saver but not if doors and windows are kept open. Cars have quite different fuel consumption in stop-and-go driving in the city as compared to cruising on the highway.

Your task will be to calculate the fuel consumption of a compact car under highway and city driving conditions based on near-realistic conditions. The mass of the car and driver is 800 kg. In city driving there are many stop signs, traffic lights, and pedestrians to avoid. On average, the car is assumed to travel 400 m between stops and it accelerates each time to 50 km/h. On the highway, the same car is assumed to travel at 100 km/h and it stops only once every 200 km. Friction is also a factor. It is assumed to be proportional to the square of the speed of the vehicle. For the purpose of this problem ignore heating, air conditioning, stereos, lights, flexing of the tires, and idling at stop signs, none of which are really negligible.

The energy efficiency rating of cars is expressed in liters of fuel used for each 100 km of travel.

a. Start by calculating the energy required to return the car to speed after a stop. This energy is wasted in friction by the brakes each time the car stops. Then calculate how many times, on average, this energy must be supplied to travel 100 km in the city and on the highway.

b. Air friction is approximated by the expression $f = cAv^2$ where c is related to the streamlining of the vehicle and A is the frontal area of the vehicle. For current model cars c can be taken as 0.32 and A as 2.1 m^2. (Vans and trucks have much larger frontal areas and less streamlining). The energy required to overcome the friction is equal to force times distance. Calculate how much energy is required for city driving and for highway driving to overcome air friction over the distance of 100 km. Neglect the fact that while accelerating the car will briefly travel at less than full speed.

c. Calculate the total energy requirement to travel 100 km in stop-and-go driving in the city and while cruising on the highway. Notice that the stop-and-go aspect plays a significant role in city driving, while air friction dominates on the highway.

d. Now take into account that the engine itself is inefficient, that only 1/7 of the energy in the fuel used actually propels the car, and that one liter of gasoline has an energy content of 37 MJ. On this basis calculate the fuel consumption of the car in the city and on the highway in liters/100 km. The published efficiency ratings for cars are in the range of 6 to 12 liters/100 km. Also be aware that automobile engines become more efficient at constant speed, particularly with high gear ratios.

1.5 MOMENTUM

1.5.1 The sport of curling is based on a number of physical principles for which expert curlers gain an intuitive feeling. The rock that the curler slides on the ice to a target area has a mass of 20.0 kg. The distance from the point of release (the hog line) to the center of the target area is 28.3 m. The coefficient of kinetic friction of the rock on the ice is 1.41×10^{-2}. The first requirement in playing the game is to get the rock as close as possible to the center of the target area, a circle of 61 cm (2 feet) radius. The opposing team tries to dislodge the rocks already in the target area by using their rocks to knock out the rocks on the ice, preferably at least 183 cm (6 feet) from the target center.

a. Determine the initial speed of the rock as it leaves the hand of the curler so that it will come to rest in the center of the target. How much time does it take for the rock to come to rest?

b. The coefficient of friction of ice depends significantly on temperature and the smoothness of the ice. What percentage variation from 1.41×10^{-2} can the curler tolerate with the same launching speed to keep the center of the rock within the radius of the inner circle?

c. Assume the first player has placed a rock at the center of the target area. The opponent tries to knock it out by a head-on collision. The first rock is to be moved 2.00 m. What speed must this first rock have gained from the second rock?

d. The collision can be considered as perfectly elastic. At what speed must the second rock be launched from the hog line to satisfy the condition of *part a*?

1.5.2 **NOTE**: This problem is completely artificial and should therefore have no place in a collection entitled ***Real-life Problems for Introductory General Physics***. However, it is fun to take some basic principles to their extremes and to see where they lead. Just imagine…

Aaron has been transported to the center of a small frozen lake. The ice is so smooth that whatever he tries he can create no friction between himself and the surface of the ice. He cannot even chip holes in the ice with his teeth. As a well-educated physics student Aaron immediately realizes how he can get to the shore. He takes off one boot and throws it as hard as he can in the direction opposite to the direction he wishes to go. The boot travels along in the direction it was aimed, bounces elastically off the shoreline, and returns to Aaron enabling him to catch it and to put it on again.

a. In imperial units Aaron weighs 112 lb fully dressed, of which his boots weigh 1.08 lb each. Clearwater Lake, where he is stuck, is perfectly round with a radius of 188 m. He throws his boot at a speed of 11.6 m/s as measured with respect to the ground. Calculate the time it will take him to reach the shore.

b. Suppose Aaron did not catch the boot after one bounce but waited to catch it until after the second bounce from the opposite shore. What would happen?

c. Try the same problem again but strictly in algebraic terms. The radius of the lake is R; Aaron's mass, fully dressed, is M; and the mass of the boot is m. The boot is thrown with a speed v.

Aaron's progress across the ice can be divided into three segments—the distance he covers in the time it takes for the boot to reach the shore; the additional distance he moves and time required for the boot to catch up with him; and finally the distance he travels with the boot back in his possession. Show that each of the three distances involved is independent of the speed with which the boot is thrown.

d. Derive expressions in terms of R, M, m, and v for the time it takes for each of the first two segments. Show how the distance and time for the last segment can be obtained.

e. If the object thrown backward were massive enough, then it is possible for it to rebound and catch up with Aaron just as he reaches the far shore. In other words, there will only be two distinct segments of the journey instead of three. Find the ratio of m to M for this to occur.

1.5.3 Cartoon characters rarely obey the laws of physics. Consider what might really happen in one situation. Mighty Mouse, in his never-ending battle with feline villains, sees a cat about to pounce on an unsuspecting mouse. The cat has a mass of 5.22 kg and approaches his victim in a horizontal direction at 12.3 m/s. Mighty Mouse comes at the cat from the opposite direction with the intent of stopping the cat. He travels at 223 m/s and has a mass of 48.1 g. Consider two possibilities: elastic and inelastic collisions.

Figure 1.4

a. In the first case, Mighty Mouse bounces back after a perfectly elastic collision with the cat. Calculate the final velocities of the cat and the mouse.

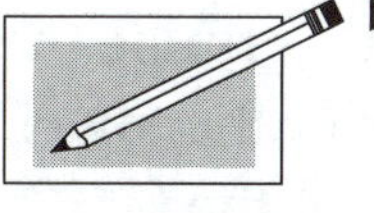

b. In the second case, Mighty Mouse has underestimated the elastic properties of the cat. He penetrates through the cat and emerges on the opposite side of the cat. He is still going in his original direction but has slowed to 44.6 m/s. Are momentum and/or energy conserved in this collision? Why?

c. Calculate the energies and momenta of the cat and the mouse before and after the second type of collision.

1.6 STATICS

1.6.1 Construction cranes for high-rise buildings are placed in the center of the building and raised floor by floor as construction proceeds. The numbers specified in this problem have not been checked with construction companies but will give an approximate idea of a realistic situation.

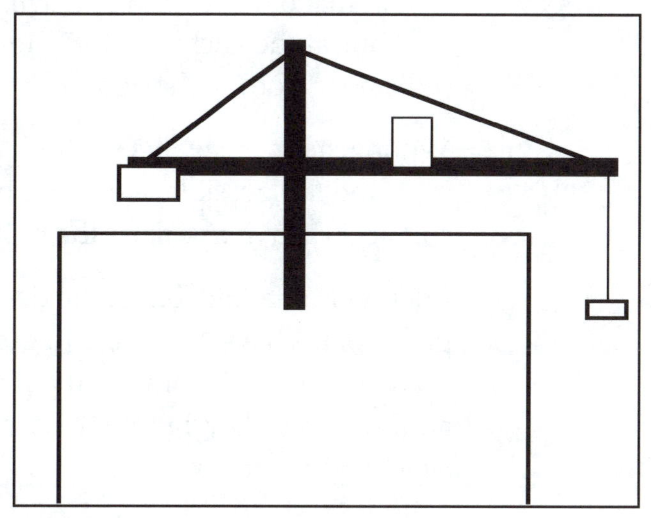

Figure 1.5

The rectangle under the left side of the crossbar of the crane represents a counterweight of 6.4×10^3 kg. The slanted lines represent steel cables to keep the two sections of the crossbar horizontal and stable. The cables extend from the top of the tower to the far ends of the crossbar. Consider these cables massless even though they are quite heavy. Real cranes also have more than one cable on each side. The top of the tower is 12 m above the crossbar and one side of each crossbar is attached to the tower by a strong hinge. On the right-hand side of the vertical tower are the cage where the operator sits and the hoist which can move to the left and right along that section of the crossbar. The cage is located halfway along its crossbar and has a mass of 300 kg, including the operator. The hoist, including mechanism, cable, and load, represents a mass of

2.5×10^3 kg. Both sections of the crossbar have a uniform mass distribution. The left side is 20 m long and has a mass of 1.4×10^3 kg. The right side is 30 m long and has a mass of 2.1×10^3 kg.

a. Each half of the crossbar must be analyzed on its own to determine the torques and forces acting to keep it in equilibrium. Consider the tower to be infinitely strong. On the assumption that the hoist is at the far right side of its section of the crossbar, calculate the tensions in the two cables and the forces acting at the hinges of the crossbar.

b. The tower is anchored in the center of the building-to-be 11 m below the crossbar. Calculate the torque of the structure about the anchor point. Recalculate the torque for the situation where the hoist and its load have been moved to directly under the operator's cage.

c. Suppose, because of uneven loading, that the tower is off the vertical by a small angle. The mass per meter length of the tower is greater than that of the crossbar. Qualitatively what will be the effect on the torque at the anchor point? How would the torque change if the distance from anchor point to crossbar were increased?

1.6.2 Yvonne, mass m_y, wants to climb a cliff of height h with the help of a rope and the possible assistance of Xantippe, mass m_x, at the top of the cliff and Zora, mass m_z, at the bottom of the cliff.

Friction is unavoidable in the described situation because all the participants have to be able to grab the rope and pull. Friction is also unavoidable as the drag of the rope along the ground or cliff face and/or at a pulley. Ignore this last aspect.

For each of the options listed below state

1) the force required,
2) the source of the force,
3) the total energy expended to get Yvonne to the top, and
4) the tension T_y in the rope holding Yvonne.

a. The rope is firmly attached at the top of the cliff. Yvonne climbs up the rope.

Figure 1.6a

Figure 1.6b

Figure 1.6c

Figure 1.6d

b. Yvonne ties herself to the lower end of the rope. Xantippe pulls her up.

c. The rope is much longer. Xantippe has fastened a pulley at the top of the cliff. Yvonne ties herself to one end of the rope and hoists herself up by pulling on the other half of the rope.

d. The pulley is still there and Yvonne is still tied to one end of the rope, but Zora pulls her up. What is the restriction on Zora's mass?

1.7 ANGULAR MOMENTUM

1.7.1 In school science competitions students are asked to build a cart that slides or rolls down a given incline faster than the cart of a competitor, subject to stringent conditions. In the real world of such a competition variables such as weight, streamlining, and lubrication are important. The idealized view of elementary physics can provide a starting point.

Four objects of equal mass M roll down a smooth surface 2.00 m long and at an incline of 7.35° with respect to the horizontal. Three of the four objects are a hoop, a disk, and a sphere, all of radius R. The fourth object is a solid rectangular block on small, essentially massless roller bearings. The densities of the objects are different. There is sufficient friction between the objects and the inclines to cause rolling without slipping.

a. Calculate the speeds with which each of the four objects reaches the bottom of the incline.

b. Calculate the time it takes each of the four objects to reach the bottom of the incline. List the four objects in order of their arrival times.

c. The three round objects roll down the incline with a uniform angular acceleration. Derive this angular acceleration from the results in *parts a* and *b*.

d. Derive expressions for the torques of the three round objects which cause the angular acceleration. These torques are to be specified using the contact point between the object and the incline as the axes of rotation.

e. Using the results of *part d* for the torque and *part c* for the angular acceleration, calculate the moments of inertia for the three round objects about the axes specified in *part d*.

f. Textbooks quote the moments of inertia for the hoop, the disk, and the sphere as calculated about their centers of mass. In *part e* you calculated the moments of inertia of these shapes about an axis displaced from the center of mass. Show that the results obtained in *part e* are in agreement with the predictions of the parallel axes theorem.

g. This problem was inspired by a parent's inquiry about the best strategy for a competition in which an object was to roll/slide down an incline and then coast an additional horizontal distance. Apply the conclusions from this problem and your own experience of the world around you to specify design characteristics for the fastest possible object of a given total volume. Keep in mind that neither the incline nor the horizontal surface that follows are frictionless.

1.7.2 A construction crane on top of a high-rise building has been dismantled for removal and only the tall center post remains standing. All but one of the bolts at the bottom have been removed. Unfortunately, the center post was not secured to the exterior crane for removal and it begins to pivot about the bolt at the bottom on the way to an unfortunate crash.

For the purposes of this problem let the center post of the crane be 24 m high, uniform in material and cross-section. Let the mass of the center post be 4.3×10^3 kg.

a. A moment after the center post begins rotating it is at an angle of 1.3° from the vertical. Calculate the torque about the pivot point. Also calculate both the angular and linear accelerations of the tip of the post.

b. A short time later the angle has increased to 35°. Recalculate the torque about the pivot point and both the angular and linear accelerations of the tip of the post.

c. Determine the angular velocity and speed of the tip of the pivoting post at 35°. How does this speed compare to the speed of the tip if it had been in free fall over the same vertical distance?

1.7.3 In games like ping pong, tennis, and billiards, giving the ball a spin has a noticeable effect on the way the ball bounces off the next surface. In nuclear physics the knowledge of the spin of particles coming from accelerators together with the measured spin of the collision products has produced most of the information on the properties of protons, electrons, quarks, and the rest of the elementary particles. In this problem you will be asked to analyze the simplest spin interaction, that of a hoop given a spin before it starts rolling along the ground.

Picture yourself holding a hoop with a mass M and a radius R. No numbers will be specified but the diameter of the hoop is such that it reaches almost to your waist and the mass is such that the hoop has a good, solid feel and will not be blown over by a slight breeze. You throw the hoop forward through the air with a speed v_i while at the same time, with a bit of wrist action, you spin it with an angular velocity ω_i. Depending on your skill, the spin may eventually cause the hoop to roll forward or backward. Also assume that there is a constant frictional force **f** between hoop and ground when the hoop slides along the ground, but that **f** is zero when the hoop is rolling.

a. Derive an expression for the initial linear and rotational kinetic energy of the hoop before any frictional losses occur.

b. Write the expressions for the initial linear and angular momentum before any frictional losses occur.

c. Under unusual conditions, the speed of rotation of the hoop exactly matches the forward speed of the hoop such that the hoop rolls without sliding as it comes into contact with the ground. At all other times the frictional force **f** between ground and hoop soon turns to friction-free rolling motion. This is now a one-dimensional problem.

Let Δt be the time it takes over which the constant frictional force acts to change the combination of rolling and sliding motion to pure rolling motion of the hoop. Symbolically calculate the change in linear and rotation momentum due to the force f.

d. State the relationships between angular and rotational momentum and energy at the time when pure rolling motion has been achieved.

e. Are energy and/or momentum conserved during the time Δt? Specify the changes that occur.

f. Now to the crux of the problem. Let the linear momentum p be positive if the direction of the center of mass is in the forward direction. Let the angular momentum L be positive if the direction of rotation by itself would lead to rolling in the forward direction. Use the information gathered in the previous parts to predict v_f, the speed of the hoop when it has reached pure rolling motion, in terms of v_i, R, and ω_i.

g. Several special cases can be selected for a detailed look from the expression derived in *part f*. What spin should be given to the hoop that it will roll without friction as soon as the hoop hits the ground? What spin should be given to make the hoop stop and fall over? What spin should be given to make the hoop come rolling back to your hand at a speed equal to v_i?

1.7.4 In Problem 1.7.3 you were asked to analyze the motion of a hoop thrown straight ahead with an additional, specified spin. The problem, with its numerous subsections, became too long and still did not explore all the ramifications. This problem assumes you have solved the previous problem.

a. Problem 1.7.3 only specified that the frictional force f is constant. The expression of v_f in that problem was independent of the magnitudes of f and Δt. Suppose you threw the hoop twice in an identical manner, once a highly polished floor and once on a rougher surface such as a cement side walk. What difference would it make to the subsequent motion of the hoop?

b. A bowling ball starts off with a speed of 7.3 m/s and zero spin as it is aimed at the pins. By the time it reaches the pins it is rolling without slipping. How fast is it going at that time?

1.7.5 **CHALLENGE:** Thoughts about physics can occur even in small cubicles with restricted seating capacity. Some of these cubicles have rolls of paper hanging from a wall bracket. This paper is rarely meant to be written on. In fact the paper is highly absorbent and perforated at regular intervals. Laboratory measurements have shown that a force of 2.5 N is required to tear the paper along a set of perforations. Repeated experiments have shown that a slow, steady pull on the end of the paper unrolls the paper to the desired length. A sharp jerk on the other hand tears the paper close to the remaining roll. The analysis of this often repeated experiment is similar to the rarely repeated demonstration analyzed in Problem 1.2.5. The student is urged to solve Problem 1.2.5 before tackling this problem.

The difficulty in tackling any non-standard problem lies in identifying the mechanisms that control the situation and the relevant variables. In this problem the assumption is that any force at the weakest points (the perforations) which exceeds 2.5 N will cause rupture at that point. Until the moment this maximum force is reached, a force from rapidly moving paper below is transmitted to the roll above. It is difficult to determine how long this force is applied and how strong it is while being applied. Careful observation of any phenomenon in which breakage occurs shows that there is a certain amount of stretch or "give" before the separation. For the paper in question this might amount to a distance of approximately 1 mm. During the time it takes to stretch the paper there will be a build-up of tension in the paper to the maximum before breakage. The tension, in turn, creates a torque on the roll and the subsequent turning motion of the roll. The final objective of this exercise is to determine the functional dependence of the turning of the roll on the vigor of the jerk, the strength of the paper at the perforations, and the mass of paper remaining on the roll.

a. Calculate the angular acceleration and tangential acceleration of the paper on the rim of the roll during the time interval that the paper stretches before tearing. Assume a constant average force 1/2 of 2.5 N and a roll of paper in the form of a uniform disk with mass 110 g and radius 6.25 cm.

b. During the time the roll accelerates at the rate calculated in *part a* the paper stretches a distance $d = 1$ mm and unwinds a distance x for a total distance $d + x$. The jerk on the paper which sets the process in motion moves the free end of the paper with a constant speed $v = 1.2$ m/s. Write an expression for, and calculate a numerical value of the time interval over which the roll of paper is accelerated.

c. Calculate the distance x the paper unwinds.

d. Show algebraically that $x = (F/m)(d/v)^2$ for $x << d$.

NOTES

2 SIMPLE HARMONIC MOTION

The world around us is filled with regularly repetitive events. Historians claim that history repeats itself. Physics, however, takes a more modest view—only repetitive events controlled through a specific form of force law can easily be described mathematically. For these simplest mathematical laws to hold, it must be assumed that whenever the system is out of equilibrium, a force arises which directs the system back to equilibrium and which gets stronger the farther the system gets from equilibrium. In fact the statement must be more specific: numerically, the restoring force must be proportional to the extent that the system is away from equilibrium. This is Hooke's Law, written algebraically as

$$\Delta \mathbf{F} = -k\,\Delta \mathbf{x}$$

where $\Delta \mathbf{F}$ represents the increase in the restoring force, $\Delta \mathbf{x}$ is the deviation from equilibrium, and k is the experimentally determined constant of proportionality between the two. The negative sign in the equation is there to remind us that the restoring force $\Delta \mathbf{F}$ acts in a direction opposite to $\Delta \mathbf{x}$, the movement from equilibrium.

The second requirement for simple harmonic motion is that once the restoring force is allowed to take over it pulls too hard, the system overshoots the equilibrium, the force reverses, and the situation repeats without end as long as we can ignore friction. The energy remains in the system.

The standard illustration of Hooke's Law is a mass hanging from a spring. When you pull the mass down, the spring stretches. When you let go, the stretched spring forces the mass upward. The mass then passes the equilibrium point and proceeds to compress the spring. As a result the spring pushes the mass back down; it again overshoots the equilibrium point, stretching the spring, and the motion repeats itself. Few systems follow these requirements perfectly, but so many systems come sufficiently close to the ideal that a familiarity with a range of phenomena where applications exist is important. There are also many regularly repetitive events in nature which do not obey these force laws. Think about going to bed every night and getting up again the next morning. There is repetition but there is no simple force law.

Two relationships common to all vibrations that obey Hooke's Law will be assumed to be experimentally or theoretically "proven." These are

$$f = \frac{1}{2\pi}\sqrt{\frac{k}{m}} \quad \text{and} \quad y = A\sin(2\pi ft + \phi).$$

In these equations f is the frequency of the oscillation in cycles per second, k is the experimental constant of proportionality, m is the "mass," y is the "distance" of the system from equilibrium at any time t, and A is the maximum "distance" from equilibrium. The symbol ϕ is known as the phase angle. Quotation marks are used in the description of m, y, and A because, in some situations, neither mass nor distance may be involved. For example, in some problems to follow, y is an angle measured in degrees instead of a distance measured in meters.

The following problems cover a wide variety of natural phenomena to which the above rules apply to a reasonable extent. In addition some problems involving repetitive phenomena are included where the rules definitely do not apply.

2.1 SIMPLE REPETITIVE MOTION

2.1.1. An angler catches a fish. To the angler, measuring the mass of the fish is almost as important as catching the fish. Suppose that the angler's pocket spring scale stretches 1.87 cm when a 1.32 kg fish hangs from it. (Figure 2.1)

a. What is the spring constant k for the scale?

b. The angler knows the relationship between mass and frequency of vibration of a spring. At what frequency should the angler expect the fish will bob up and down (oscillate) while suspended from the spring?

c. By what additional distance will the spring stretch if a second fish with a mass of 538 g is added? Calculate the frequency at which the two fish will oscillate on the spring.

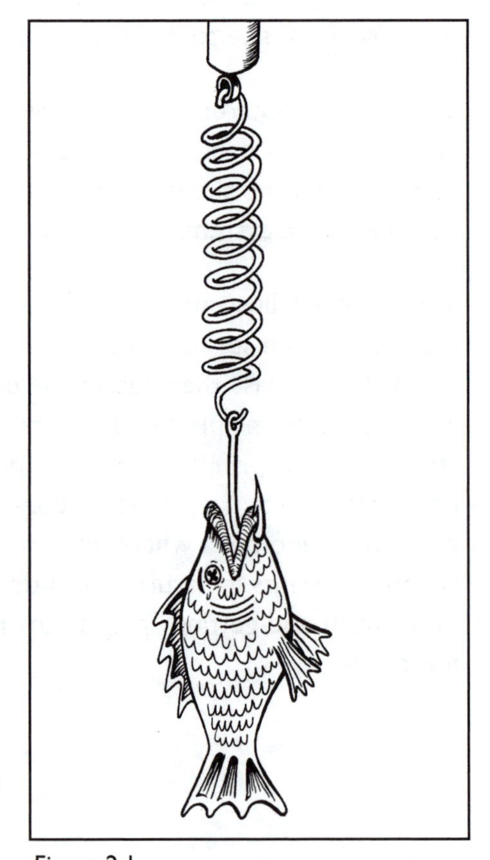

Figure 2.1

d. Since f is proportional to $1 / \sqrt{m}$ why does a single, simple unloaded spring not vibrate at an infinite frequency?

2.1.2. A heavy automobile of the 1970s vintage has a mass of 1700 kg. Since cars have four wheels, they also have four independent springs, and the weight of the car is evenly distributed over all the springs. The car is lightly loaded with three adults and their luggage. As a result the car sinks 1.5 cm on the combined springs. Passengers and luggage together have a mass of 310 kg. The car with its load is slowly driven over a rough road. As a result the passengers may suffer motion sickness from its up and down oscillation.

a. What is the effective spring constant of the car's suspension? Calculate the frequency at which the car 's chassis (passengers included) will oscillate.

b. A more modern, energy-efficient car has a mass of 700 kg and is capable of carrying the same number of passengers and the same amount of luggage as the car described above. It also has springs so that it will not scrape the ground when loaded. Assume then that the car sinks the same 1.5 cm under the same load of 310 kg. Calculate the effective spring constant of this car.

c. At what frequency will the more modern car and its load oscillate? Will this car be considered to have a "smoother" or a "rougher" ride than the 1970s model?
Why?

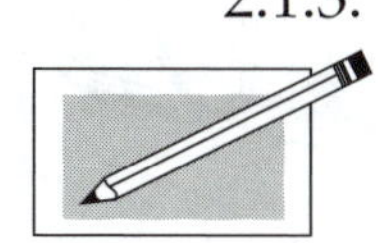

2.1.3. Maxine and George like to go to a playground where there are large vertical springs topped with animal figures forming a child's seat. The ride for the children consists of bouncing up, down, forward, and sideways on the spring. Whichever rides each child tries, George consistently gets a bouncier ride than Maxine. Occasionally their mother tries the ride, but hardly bounces at all. Figure 2.2 shows just Maxine on the ride. Explain why each person gets a different ride. Age appears to play a role. Is this so? Why or why not?

Figure 2.2

2.1.4. Weigh scales often use springs for weight or mass determinations because they are small and cheap. A mass suspended from a spring stretches it by a predictable amount. More accurate weigh scales are made like balances. For each type of scale the calibration can depend on the local value of the acceleration of gravity.

a. If gravity were reduced as it is on the moon or in outer space, explain how that would affect the results when a spring scale is used as a measuring tool for weight and for mass.

b. A balance scale could also be taken to the moon or into space to compare masses and/or weights. Explain how the function of this device would be affected if gravity were greatly reduced.

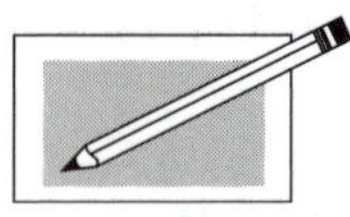

c. Imagine hanging a mass M from a spring, here or on the moon. Set it into oscillation. Measure the frequency or period of the oscillation. Does the frequency depend on gravity? Does M depend on gravity? Does k depend on gravity? Justify your response.

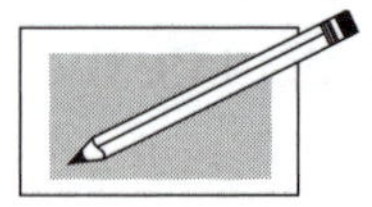

2.1.5. Some mattresses are made by sewing individual coil springs into side-by-side pockets over the area of the mattress as shown in Figure 2.3. The *Beautyrest* mattress is of this type. Other types of mattresses, those from fold-away beds in particular, have springs and metal straps tied together. Describe in a brief essay how these different types of sleeping platforms respond to the user. Include in your discussion what happens when a small child jumps on the bed.

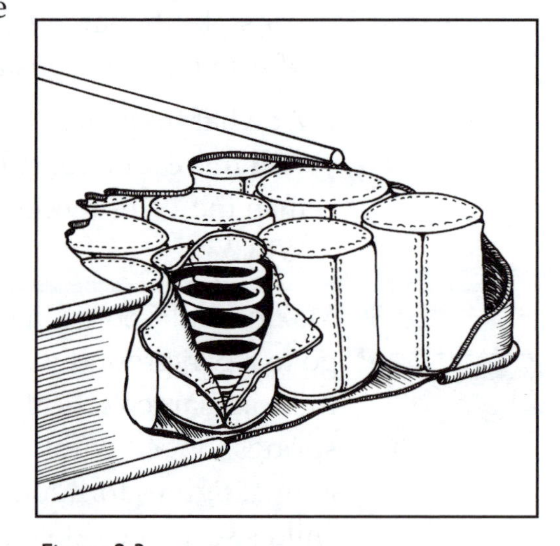

Figure 2.3

2.1.6. In mechanically oscillating systems a mass swings in and out of equilibrium. It is pulled back by a spring or gravity. Electrical systems can also oscillate. Two distinct circuit elements are needed for electrical oscillations to occur. These are a capacitor C and an inductor L. The oscillations occur when the electrical energy in the system oscillates back and forth between the capacitor and the inductor. Such circuits are found in all radios.

The mathematical expression for the frequency of the oscillation is

$$f = \frac{1}{2\pi\sqrt{LC}}$$

and the SI units for the inductor L is the henry (H) and for the capacitor C is the farad (F). Both units are impracticably large, and inductors and capacitors as circuit elements are usually in the micro henry (μH) or pico farad (pF) range. When an external electrical oscillation of a certain frequency is brought close to a circuit which can oscillate at that same frequency, the circuit will pick some of the external energy for its own oscillations. In practice the antenna of a radio picks up a broadcast signal from a remote radio station and feeds the small amount of energy received to an L C combination. Since the standard (AM) broadcasting range is from 530 kHz to 1600 kHz (kiloHerz), the L C combination must be tuned to the frequency desired station. The capacitor C is an element that can be easily varied. The inductor L does not change in tuning the radio.

a. If the value of L is 250 μH for a certain radio, what must the value of the capacitor be to tune into (receive) a station which broadcasts at 550 kHz?

b. What must the value of the capacitor be to receive a station which broadcasts at 1600 kHz at the opposite end of the dial?

c. For FM reception the required frequency range is from 110 MHz to 88 MHz (MegaHerz). For this application one also uses an L C combination with a variable capacitor and a constant inductor to tune into stations in that frequency range. If the inductor in this case is 0.50 μH, calculate the range over which the capacitor must be variable to tune over the entire FM band.

2.1.7. Much of the progress of modern chemistry depends on knowing the strengths of the bonds and the location of atoms in complex molecules. Under electromagnetic stimulation the molecules can be set into vibration in many characteristic modes. The frequency of each of these modes depends on the strength of the local bonds (the restoring forces) and the masses of the atoms in the immediate vicinity. To a first approximation these systems obey Hooke's law.

One way of assigning a specific bond strength to a specific location is to use isotope substitution. A slightly heavier or lighter atom of the same element is substituted at the molecular site to be probed. The result is a shift in the vibrational frequency results. The most striking results are in molecules which contain hydrogen because a deuterium atom (heavy hydrogen, mass = 2 atomic mass units (amu)) can be substituted for the ordinary hydrogen atom (mass = 1 amu).

a. Hydrofluoric acid (HF) is one of the most corrosive liquids used in the laboratory. The HF molecule is a simple one in which the hydrogen atom is bonded to the much heavier fluorine atom. The combination can be brought into vibration. The characteristic frequency of the atoms in this bond is 1.24070×10^{14} Hz. Deuterium, at twice the mass of the ordinary hydrogen atom, can be substituted for the hydrogen atom to make DF. Based on the (oversimplified) model of the mass on the spring, predict what the vibrational frequency of the DF molecule should be. (Experimentally that frequency is found to be 8.98853×10^{13} Hz.)

b. The disagreement between simple theory and experiment as obtained in *part a* is too large for both to be correct. A simple correction can be made to the calculations in *part a,* because the fact that the fluorine atom has less than an infinite mass was ignored. For two masses m_1 and m_2 to vibrate freely relative to each other on a spring, it is necessary to substitute the reduced mass μ defined as

$$\mu = \frac{m_1 m_2}{m_1 + m_2}$$

for the single mass m which appears in the equation for the frequency of the mass oscillating on the spring. The constant k is not affected. Calculate the reduced masses to be used in frequency calculations for HF and DF based on F having a mass of 18.99840, H having a mass of 1.007825, and D having a mass of 2.01410. Then, using the known HF frequency, make a new prediction of the DF frequency.

c. The reduced masses allowed you to calculate the vibrational frequency of DF knowing the vibrational frequency of HF. Again assuming the system simply acts like two masses connected together with a spring, calculate the force constants for the HF and DF bonds. (Hint: convert the reduced masses from atomic mass numbers (amu) to kilograms (MKS units). The important conversion factor is that 1.00000 amu is equal to 1.67262×10^{-27} kg.)

2.1.8. The chlorine gas molecule is Cl_2. There are two stable isotopes of chlorine: ^{35}Cl (mass = 34.96885 amu) and ^{37}Cl (mass = 36.96590 amu). Therefore there can be three distinct Cl_2 molecules: $^{35}Cl_2$, $^{37}Cl_2$, and $^{35}Cl\,^{37}Cl$.

a. The vibrational frequency of $^{35}Cl_2$ is 16.779 x 10^{12} Hz. Using the concept of reduced mass (see Problem 2.1.7b), predict the other vibrational frequencies found in chlorine gas.

b. Molecules can be identified and their relative concentrations can be measured in mixtures of gases by noting the characteristic frequencies absorbed by the gas and by measuring how strongly a given characteristic frequency is absorbed. For a given gas the absorption is proportional to the amount of that gas present in the mixture. Given that about 75% of Cl atoms are of the ^{35}Cl isotope and about 25% are of the ^{37}Cl isotope, predict the relative strengths of the absorptions of the three possible Cl_2 molecules.

2.1.9. Restaurants can be interesting places to study simple harmonic motion. In particular it can seen there that certain aspects of circular motion show more than coincidental similarities to simple harmonic motion. Some restaurants serve several varieties of dishes on a rotating stand anchored to the center of a round table. Imagine just a single pot of jasmine tea on the edge of such a slowly revolving (3.0 rotations per minute) food server. Albert, Betty, and Chan are sitting around the table 120° apart watching the teapot. Also imagine a bright light behind each of the diners. Screens are placed on the opposite side of the table from each light where a clear shadow of the teapot can be seen at all times. Figure 2.4 illustrates the situation. The motion of the shadow of the teapot on the screens is the subject of this problem.

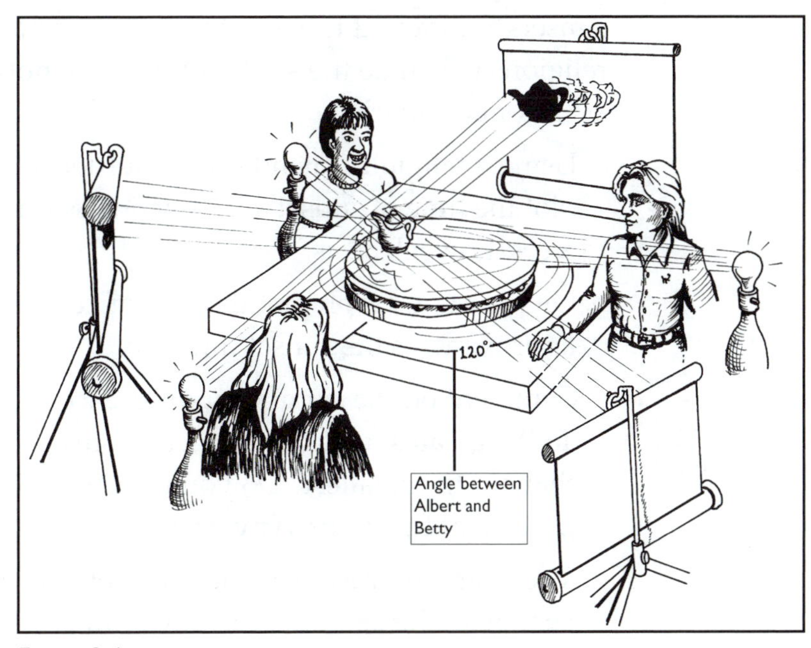

Figure 2.4

The diners all have digital watches with a stopwatch feature. They all start their watches at a signal given by Albert. Albert gives the signal at the moment when he sees the shadow of the teapot at its farthest point to the right on the screen. For all observers, each looking at the screen opposite them, the shadow travels back and forth over a distance of 2.5 m. Figure 2.4 shows the seating arrangement, screens, and direction of the rotation of the food service tray.

NOTE: Even though the teapot is actually revolving in a circle on the rotating food tray, its shadow only moves back and forth on the screen.

a. Write a mathematically precise expression for the motion of the shadow as seen by Albert on the screen directly across from him.

b. Write mathematically precise expressions for the motion of the shadow as seen by Betty and by Chan on the screens directly across from each of them.

c. The rotating food service tray has reasonably frictionless bearings. As a result it continues to revolve without any obvious forces being applied. The shadows move freely on the screen because they have no mass. The teapot, however, does have mass. Show that the force on the teapot does satisfy the requirements that go with simple harmonic motion. Consider the consequences if Betty decides to increase the spin of the rotating food tray.

2.1.10. As the year passes, the amount of daylight changes from short times in the winter to longer times in the summer and back again, in what looks like simple harmonic motion. Knowing the exact times for sunrise and sunsets at different times of the year is of particular importance in religions which tie the start and finish of holy days or months to this astronomical marker.

a. Using a religious calendar for your locale, or the daily newspaper, plot the time of sunset as a function of date for a complete year on a week-to-week basis.

b. On the same graph plot the sine or cosine function which most nearly approximates the sunset function.

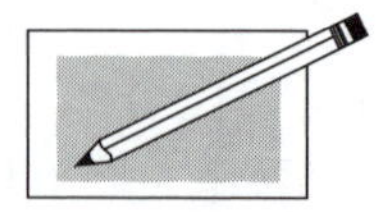

c. Comment on the significant differences between the two plots. When do the greatest week-to-week changes take place? When are these changes the smallest? Do you expect the length of the day to be a sinusoidal function? Why or why not?

d. The length of the day is the time between sunrise and sunset. A change in the length of the day in the evening is closely, but not precisely, matched by a change in the length of the day in the morning. That is, as the sun sets later in the evening, it also rises earlier in the morning. Sketch a graph of the length of the day as a function of date for your location.

e. There are two extremes of the yearly day and night cycles which occur on the earth at the poles (North and South) and at the equator. Sketch a graph of the length of the day as a function of date for these three locations. Explain why it should be so different from your locality.

2.1.11. Birds migrate every year. The Arctic tern travels farther than any other bird, flying from Arctic to Antarctic to spend the summers at opposite ends of the globe. In North America birds such as pelicans spend the summer breeding in northern Alberta and Saskatchewan and spend the winters along the coast of California.

a. Sketch a graph of position (north and south) vs time of year for a migrating bird species of your choice.

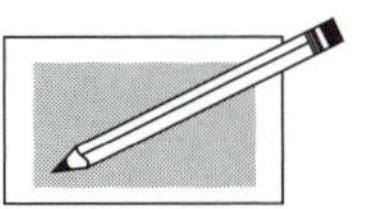

b. Can the migration patterns be considered simple harmonic motion? What force is operating in this situation? Describe the ways migration patterns do or do not correspond to Hooke's Law.

2.1.12. Humans breathe in a reasonably regular fashion. Air entering and leaving the mouth and the nose is one characteristic of the respiratory process; the rising and falling of the chest is another. Write a brief report in which you compare this biological repetitive phenomenon with simple harmonic motion. Observe your own breathing and consider the forces involved.

2.2 ENERGY CONSIDERATIONS IN SIMPLE HARMONIC MOTION

Energy loss in all motion is a fact of life. Automobiles need fuel to move at constant speed because of friction. Children on swings stop swinging unless they are regularly given a push or unless they have learned to pump for themselves. Energy loss is even deliberately introduced in mechanical systems to reduce vibration. An automobile on springs but without shock absorbers (creating friction against vibrations) would be extremely uncomfortable.

The following problems examine, in a simplified way, how energy can be provided to or lost from oscillations. One specific fact is common to all the problems—the total energy in any oscillating system is proportional to A^2, the square of the maximum amplitude of the oscillation.

2.2.1 A child is happily swinging back and forth having been pushed to swing through a large arc. Because of friction at the swing supports and the wind that is created, the arc of the swing steadily decreases. An approximate description for the energy loss in a system like this is that a certain percentage of the total energy of the child on the swing is lost after each complete (back and forth) swing.

Deborah sits on a swing that takes her through arcs of 3 m radius. She swings through a total angle of about 60° with a period of 3.5 s. Suppose that the starting energy for Deborah on the swing is 75 J and the starting height of the swing is 0.40 m above its lowest level. The energy loss per swing is 9.0%.

a. Plot two curves on the same graph. Plot one curve of the total energy in the swing as a function of time. Plot a second curve of the amplitude of the swing as a function of time.

b. After how many swings will the energy of the system have decreased to below half its original value? After how many swings will the amplitude have decreased to below half its original value?

The timing mechanisms in mechanical clocks are based on a pendulum or a coil spring. Modern digital watches have a vibrating quartz crystal for this purpose. In each case there is a fundamental frequency of oscillation which has to be kept going for days or years at a time. Each of these mechanisms requires compensation for friction in the form of a brief but regular push to keep the amplitude of the oscillation at a near constant level. The same principle is true for the parent who keeps a child going on a swing with a push each time the child reaches the backward arc.

2.2.2 If you have had the experience of pushing a small child on a swing, then you (the external agent) replace the energy lost through friction by giving the child a quick push once during each cycle.

The length of a swing is 2.9 m, the mass of the child is 15 kg. In the beginning Edward is swinging over a total arc of 70° with an energy of 77.1 J. The period of oscillation is 3.4 s. The swing loses 7.0% of its energy during each complete swing. Each time you give Edward a push you inject 2.3 J of energy into the system.

a. Calculate the equilibrium energy of Edward on the swing.

b. Calculate the amplitude of the swing under equilibrium conditions.

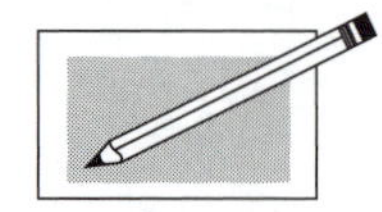

c. Explain what would have happened if, with the same regular push, the child on the swing had started from rest?

2.2.3 Ocean tides are caused by the constantly changing gravitational attractions of the moon and the sun on the Earth. On an idealized and imaginary earth with a constant depth of ocean everywhere and no dry land, there is a rise (bulge) in the water directly below the moon and a second bulge on the side opposite to the moon. A smaller bulge travels along the earth directly below the sun and on the side opposite to the sun. Because the moon and the sun are not in synchronization, the timing of the tides will shift from day to day, and sometimes the attractions of the moon and the sun will add together. The Bay of Fundy has the highest tides in the world, with as much as 17 m difference between high and low water levels when the sun and the moon pull in the same direction at the same time. Unusually high tides occur in a bay or inlet where the natural frequency of the water flowing back and forth is the same as the reappearance of the two peaks of high water due to the moon's gravitational pull. The new moon, although not visible, is at its highest point at noon local time, when the sun is also at its highest point. This creates the highest tides. Each successive day the moon reaches its highest point almost an hour later. In 28 days the moon is back to new moon status and pulls together with the sun.

a. Calculate more precisely how much later the moon reaches its highest point each day.

b. Predict what the natural frequency of the tides in the Bay of Fundy should be.

In mechanical clocks and watches the oscillation itself triggers the push at the right time during each swing to keep the system going. The energy for the push comes from a loaded spring which is wound regularly. In quartz clocks the push is an electrical one. There are situations of importance in which the push is not sufficiently well synchronized with the natural frequency. A radio which is "off station" is one example. A mechanical and an electrical example follow.

2.2.4 Frieda is on a swing this time. At the beginning the swing goes through a total arc of 50°; the effective length of the swing is 2.4 m and Frieda's mass is 18 kg. For the purpose of this problem assume that no energy is lost to friction.

a. What is the period of the oscillation? What is the starting energy of Frieda on the swing?

b. A playground supervisor stands behind the swing and gives Frieda a push at an interval which is 5% longer than the natural period of Frieda on the swing. As long as Frieda is still going forward when the push occurs, 1.7 J is added to the energy of the swinging child. After a while the push occurs while Frieda swings backward rather than forward, thereby effectively braking the swing and removing 1.7 J from the total energy each time.

Calculate how many successive pushes can be given that add to the energy of the swing under these circumstances. How much energy is added? What will become the maximum height during the swing?

c. What will be the maximum height during the swing after the complete set of pushes in which energy is removed from the swing?

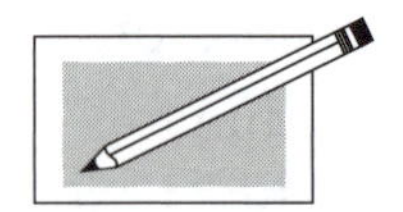

d. Explain what will happen as the difference between the time interval of the pushes and the natural period of the swing come closer together. The frequency at which the cycle of maxima and minima in the energy in the swing occurs is called the beat frequency.

The broadcast and reception of music from a radio station is analogous to the mechanical operation of the child on the swing and the supervisor giving a push. The differences are in the timing and in the form of the push. The child on a swing has a frequency below one herz (Hz)—a period between one and two seconds. In radio the resonant frequency (the station you tune in to) operates at a frequency near one megaherz (MHz), or with a period of about one microsecond. The mechanical push that keeps the swing going is replaced by the electrical energy from the radio station. When the station sends out a steady signal, the receiving circuit in the radio swings at a steady amplitude. When the signal from the station stops, the receiving circuit will decrease its amplitude as dictated by the dissipation of the energy in the circuit; just like the child on the swing will slow due to friction. Neither the circuit nor the swinging child stops immediately after the energy input ceases.

Figure 2.5a shows the amplitude of the radio signal in the receiver under steady broadcast power. Figure 2.5b shows what happens when the station suddenly goes off the air. Figure 2.5c shows what could occur in the receiver when the station regularly increases and decreases its power output. A line connecting the peaks in Figure 2.5d could be a higher frequency wave which would become a higher frequency note in the loudspeakers of the radio.

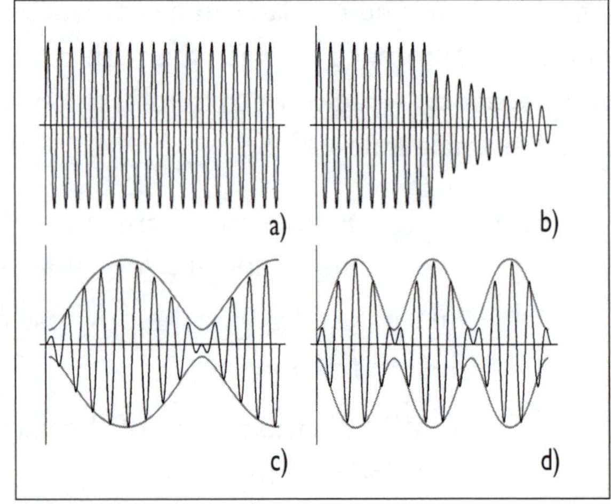

Figure 2.5 a, b, c, & d

A few principles are important to keep in mind when dealing with oscillating systems that are supplied with energy. The first is the principle of tuning. For a radio to operate and be tuned to a station, the radio circuit must be able to store energy at the frequency of the station. A child's swing is not much fun for either party if it stops as soon as no one pushes. The second principle is that the less energy is lost between swings the more efficient the system becomes in accumulating energy from the source. The third principle is that trouble arrives when the system becomes too efficient, that is, if the losses per swing are too small. The system can no longer respond quickly to changes, and high frequency signals are washed out. The musical implication is a poor high-frequency response. In mathematical terms the decrease in maximum amplitude is closely approximated by an exponential decay in the form

$$y = y_0 e^{-\alpha t}$$

where y is the amplitude of the wave in the circuit at a time t after the power from the station has been shut off, y_0 is the amplitude in the circuit before the power is shut off, and α is the number which characterizes the rate at which the amplitude drops.

2.2.5 A local AM radio station broadcasts at a frequency of 880 kHz. The energy per second reaching the receiving circuit of a radio from the station varies from 0 μW to 0.15 μW depending on the volume of the music being broadcast. To be able to respond to the highest audible frequency at 20 kHz, the receiver must have enough dissipation for the 880 kHz oscillation to follow a rise and fall at 20 kHz. AM stations actually don't send signals of such high frequencies; 5 kHz is a practical cut-off.

a. For a minimal musical response we should expect a drop in amplitude of the 880 kHz radio signal of at least a factor of 100 at 10 kHz. Calculate the time interval from peak to valley of a 10 kHz oscillation. In that time the 880 kHz oscillation in the circuit must drop to 1/100th of its maximum value. Calculate the dissipation constant α required.

b. By what fraction will the amplitude of the 880 kHz signal decrease for each period of that oscillation? Energy is proportional to the square of the amplitude. What fraction of the energy will be lost at each cycle of the 880 kHz oscillation?

c. Calculate the energy stored in the receiving circuit of the radio as a result of a steady input of 0.15 μW at 880 kHz from the radio station.

d. Hearing in the high-frequency range deteriorates with age. Most older people can still hear 5 kHz audio signals. Use the value of α you have calculated in *part a* to determine the drop, from highest to lowest amplitude, that the radio should be able to reproduce at that frequency.

2.2.6 The human heartbeat is relatively regular and repetitive. It can be heard through a stethoscope or felt at the pulse.

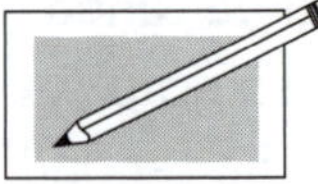

a. Explain, in as much detail as possible, why the motion you detect with the stethoscope or at the pulse is NOT simple harmonic motion. Specifically comment on the shape of the blood pressure vs time curve and restoring force.

b. Some people need pacemakers to keep the heart beating at a regular rate. Does the pacemaker supply energy to the system in a way similar to the person pushing a child on a swing?

Explain what physiological problems might exist when a person with a pacemaker exercises and needs to raise his heart rate.

3 FLUIDS

3.1 BUOYANCY

3.1.1 A traveler in China will see barges made from cement on the rivers and canals. At first glance one would think that these barges should sink like a rocks. In fact they are practical where wood and steel are scarce or expensive even though they may crack more easily than their wood or metal counterparts. The subject of this problem is the carrying capacity of a cement barge of simplified shape.

Imagine a barge shaped like a rectangular box, open at the top. The outside dimensions are 5.0 m wide, 20 m long, and 2.0 m deep (the height of the sides). The sides and the bottom are made of cement 5.0 cm thick, with a specific gravity 2.9.

a. How deep will this barge be immersed in the water when it is completely empty?

b. In absolutely still water how much mass, in kg, can the barge hold without sinking?

c. A typical load to be transported in China might be cabbages. The density of a single cabbage is slightly less than the density of water because of the spaces between the cabbage leaves. A load of cabbages will have additional air spaces between the individual cabbages. The specific gravity of a load of cabbage might be about 0.7. Determine the volume of cabbage the barge could carry in still water. Is it more or less than the volume of the barge?

3.1.2 Weather observation balloons and dirigibles are most often filled with helium even though hydrogen is cheaper and lighter. Here the flotation of balloons is investigated and hopefully will answer the question: If helium makes a balloon go up, then why not compress the helium and make it rise even more?

The buoyant force a fluid exerts on an object is the difference between the force of gravity on the object and the force of gravity on the fluid displaced by the object. Here the fluid is the air and the object is the balloon with the load it carries.

The aim of this problem is to calculate the requirements for a balloon to lift (i.e. support off the ground) a total load of 800 kg. This load is made up of the basket, several passengers, supplies, sandbags for ballast, ropes, and the skin of the balloon itself. The 800 kg does not include the gas fill of the balloon. The volume occupied by the load is, in the first instance, negligibly small compared to the volume of the balloon itself. Imagine then the balloon plus load to be a sphere.

a. Calculate the density of atmospheric air at 273°K based on the mass 28.8 grams for one mole of air occupying 22.3 liters at one atmosphere. (Air is a mixture of approximately 80% nitrogen, 20% oxygen)

b. Suppose it was possible to make a large sphere with a thin outer shell capable of sustaining a vacuum inside, to act as a balloon. Determine the size of the sphere, first volume and then the radius, needed to lift this balloon. Remember that the 800 kg load includes the weight of the shell but not the gas fill.

c. It is practically impossible to make a lightweight shell that can sustain the pressure difference of one atmosphere on the outside against zero atmospheres on the inside. Therefore, the inside of the balloon is filled with a light weight gas to a pressure equal to the outside pressure. At standard temperature and pressure, one mole of helium has a mass of 4.00 grams and a mole of hydrogen gas has a mass of 2.00 grams. Determine the size of the sphere, first volume and then the radius, needed to lift this balloon when filled with helium and when filled with hydrogen.

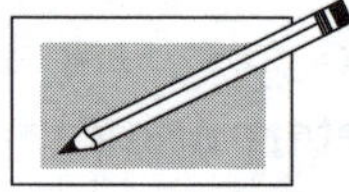

d. Explain what would happen if the pressure of the fill gas within the shell of the balloon is doubled, assuming the shell is strong enough.

3.1.3 Swimmers, snorkelers, and scuba divers notice a significant difference between their activity in fresh water and in sea water. The difference becomes even greater if they tried swimming in the Great Salt Lake or the Dead Sea. The difference is due to the specific gravity of the water: 1.000 for fresh water (H_2O), 1.025 for sea water (CH_2O?), and 1.20 for salt saturated lakes like the Dead Sea. The specific gravity of lead is 11.3.

For this problem it is useful to know that to a good approximation one liter of a substance has a mass in kilograms numerically equal the specific gravity of that substance.

The average specific gravity for a human being varies from person to person. It depends on bone structure, fat content, and breathing. For the purposes of this problem, assume that Tarzan has an overall specific gravity of 0.990 and a mass of 82.8 kg. Also assume that the human head, Tarzan's included, occupies a volume of 3.10 liters and has a slightly higher density than the rest of the body.

a. Tarzan quietly floats in fresh water in the company of alligators. He is just barely breathing and only a part of his head shows above the surface. What percentage of his anatomy will be above the water? Does that depend on which part of his anatomy he keeps above the surface? Explain.

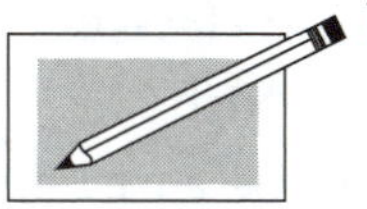

b. Tarzan tries floating in fresh water in two different orientations. First he tries floating on his back, with the back of his head firmly under water. Then he tries floating with his body vertical and part of his head showing above the surface. Explain why he can breathe freely in the first case, but will need a breathing tube, like a snorkel, to get air when floating vertically.

c. Using the same orientations as in *part b*, Tarzan is floating along the shore of the Indian Ocean off the coast of Africa in the company of crocodiles. What percentage of his anatomy is above the surface of the water in this environment? Will he be able to breathe while floating vertically? Ignore the appetite of crocodiles for human flesh.

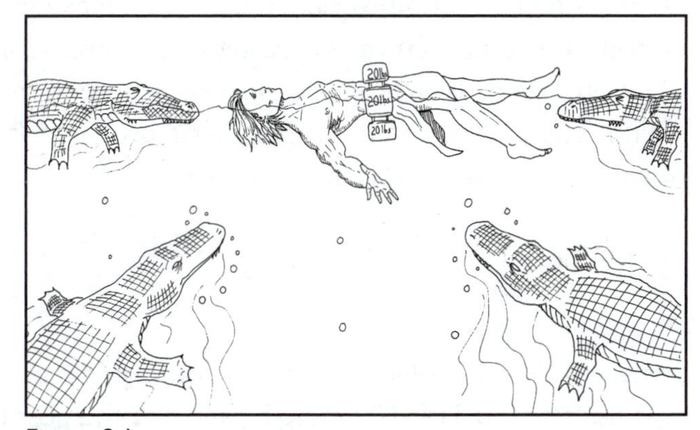

Figure 3.1

d. The nose and mouth can be kept above the water with swimming motions of the hands or feet. What constant force must Tarzan apply in fresh water to stay as high as he would if he were to hold perfectly still in sea water?

e. Scuba divers wear a belt with lead weights to achieve neutral buoyancy, i. e. a specific gravity equal to that of the surrounding fluid. They also carry compressed air bottles and other gear which Tarzan does not have available. How many kilograms of lead must Tarzan wear to achieve neutral buoyancy in fresh water, sea water, and Dead Sea water?

3.2 STATIC PRESSURE

Pressure changes with height. This is readily experienced as pain in the ear on a rapidly moving elevator or while diving under water. For all fluids the expression for the change in pressure ΔP

$$\Delta P = -\rho g \, \Delta h$$

is valid as long as the gravitational attraction g and the density ρ remain constant within the increase in height Δh. Water increases its density by about 50 parts per million when the pressure changes from one atmosphere to two atmospheres. The acceleration of gravity near the surface of the Earth decreases by 1.5×10^{-6} m/s^2 for each meter rise. Both changes are quite negligible for most calculations done to determine the pressures below the surface of the water. The same conditions hold for other common liquids, and the valid, although approximate, expression for pressure at a depth h, as measured downward from the surface, becomes

$$P = P_o + \rho g h$$

where P_o is the pressure at the surface of the liquid.

A gas is a highly compressible fluid. An increase in pressure from one to two atmospheres, at a constant temperature, will result in an increase in the density of the gas by a factor of two. The observed changes in pressure with altitude must be calculated with a changing density in mind. The expression for atmospheric pressure as a function of altitude h can be written as

$$\ln P = \ln P_o - 10^{-3}\frac{Mgh}{RT} \quad \text{or} \quad \ln\left(\frac{P}{P_o}\right) = -10^{-3}\frac{Mgh}{RT}$$

where M is the molecular mass in grams per mole of the gas in question; R is the universal gas constant, 8.314510 J/mole K; and T is the absolute temperature of the gas in kelvins. One of the assumptions made in the derivation of the above expression is that the temperature does not change with height. In fact it does, but this change is small compared to the change in density with height and is to be neglected in the following problems.

In SI units pressure is measured in pascals. The pascal (Pa) is defined as a N/m^2

3.2.1 A serious problem for scuba divers is a medical condition called "the bends." Nitrogen gas under sufficiently high pressure, such as is experienced in deep-sea diving, dissolves in the blood. When the diver returns to the surface in a controlled way, the nitrogen diffuses out of the blood without doing damage. However, if the diver surfaces rapidly the nitrogen in the blood forms bubbles, with painful and damaging results. The depth below which precautions must be taken is 35 feet (10.7 m). The maximum safe time at this depth is 205 minutes. At twice this depth dangerous levels of nitrogen are accumulated in 40 minutes.

Calculate the pressure at this depth of 70 feet.

NOTE: Helium can be substituted for nitrogen in breathing mixtures to permit deeper dives and more rapid surfacing.

3.2.2 The Dead Sea, located on the border between Jordan and Israel, is said to be the lowest point on the surface of the Earth. Its surface is at 409 m below sea level. The atmospheric pressure there is greater than at sea level. The atmospheric pressure at sea level is 101.3 kilopascals (kPa). The density of air at this pressure is 1.29 kg/m^3 when the temperature is 0°C.

a. A depth of 409 m below sea level is small on a global scale; therefore assume that the density of the air does not vary with altitude. Predict on that basis what the atmospheric pressure at the surface of the Dead Sea might be on an extremely unlikely cold winter day of 0°C.

b. Pressure increases much more rapidly under the surface of the water. How deep does Jennifer have to dive in a freshwater swimming pool at sea level to be under the same pressure as she would experience in air at the surface of the Dead Sea?

3.2.3 One of the serious difficulties in climbing high mountains is the decrease in air pressure and the associated lack of oxygen. The peak of Mount Everest is 8880 m above sea level. It is said that the air pressure there is approximately 1/3 that at sea level.

a. Ignoring reality, assume that the density of air does not change as the altitude changes. If that were the case what would be the calculated pressure at the top of Mount Everest?

b. Calculate a more realistic value of the air pressure at the top of Mount Everest using the expression which allows for the variation in the density with height. Since air is a mixture of oxygen and nitrogen it makes sense to use an average value of the molecular weight of air—28.8 grams/mole. Assume the temperature to be 0°C (273°K).

3.3 FLUID FLOW

Bernoulli's equation adds another variable to the pressure-depth relationship. In a situation where the fluid is incompressible and the flow is frictionless or where the fluid is "almost" incompressible and the flow is "almost" frictionless, one can write

$$P_1 + \rho_1 g h_1 + \frac{1}{2}\rho_1 v_1^2 = P_2 + \rho_2 g h_2 + \frac{1}{2}\rho_2 v_2^2.$$

In this equation the subscripts 1 and 2 refer to the pressures, densities, depths, and flow velocities at two different positions in the fluid.

3.3.1 The deepest known part of the ocean is the Challenger Deep which is in the Mariana Trench of the Pacific Ocean near the island of Guam. The measured depth of the water there is 11 km. Seawater has a specific gravity of 1.025 and a bulk modulus of approximately 2.2×10^9 Pa.

a. Ignoring any possible changes in salinity, temperature and density with depth, calculate an approximate value of the pressure at this deepest point of the ocean.

b. Based on the pressure calculated in part a, calculate the mass of $1.00 m^3$ of sea water at the bottom of the Challenger Deep as compared to its mass at the ocean's surface.

c. Specially designed exploration vessels have been constructed to explore at these depths. Suppose that one of these vessels develops a small leak at this ultimate depth of 11 km and fills with water under high pressure. The vessel is rapidly brought to the surface and the water sprays out of the leak. Calculate the velocity at which the water leaves the vessel. Why would this situation be considered dangerous?

3.4 VISCOUS FLOW

Liquid helium below 2.14 K is the only fluid known to science which has zero viscosity and will therefore flow equally well through narrow pores as through wider tubes. Poiseuille's Law predicts how viscosity and the radius of a cylindrical tube affect the flow of a liquid through a tube under non-turbulent conditions. Poiseuille's Law is stated as

$$Q = \frac{\pi R^4 (P_2 - P_1)}{8\eta L}$$

where Q is the fluid flow rate in m^3/s, R is the radius of the tube, P_2 and P_1 are the pressures at the beginning and the end of the tube respectively, η is the viscosity of the fluid, and L is the length of the tube.

3.4.1 Blood flows through large arteries and veins but it also flows through minute capillaries. Biological systems are complex, but for the purposes of getting an idea of how such systems work it is easier to think in terms of simple mechanical models.

a. Suppose that the aorta is represented by a cylindrical tube 1.0 cm in diameter and length L. How many tubes of 0.20 mm diameter and length L, laid side by side, would be required to support the same flow rate of the same fluid for the same pressure difference?

b. Determine the required total cross sectional area of the smaller tubes as compared to the single larger tube in order to support the same total flow under identical conditions.

NOTE: All our blood flows through the aorta, but that same total flow is also pushed through capillaries, whose diameter is the size of a red blood cell. Admittedly the pressure drop across the capillaries is larger than the pressure drop across the aorta, but the number of capillaries required is immense.

3.4.2 The water pipe entering a house has an inside diameter (ID) of 4.0 cm. A plumber must extend the line to be able to connect the sinks, shower, toilet, and washing machine. Initially the plumber uses a 4.0 cm pipe to match the outside line. After he uses his total supply of 4.0 cm ID pipe, he substitutes 4 pipes 2.0 cm ID in parallel.

a. If 2.0 ℓ/s (liters per second) of water flow through the larger diameter pipe, what will be the flow rate through each of the narrower pipes?

b. At what speed, in m/s (meters per second), will the liquid flow in both the wide and the narrow pipes?

c. Ignore friction in the pipes and calculate the required absolute pressure at the inlet side of the large pipe to achieve the 2 ℓ/s flow speed if the pipes are all horizontal and open to the atmosphere at the outlet side.

d. The viscosity of water at $\eta = 1.4 \times 10^{-3}$ Pa·s (pascal seconds) is small but not negligible. What pressure difference is required to achieve a flow rate of 2.0 ℓ/s over a 12 m horizontal length of the large diameter pipe? What pressure difference is required for the same total flow rate through the four narrower parallel pipes over the same distance?

3.4.3 Enbridge Pipelines Inc. is a utility that uses pipelines to transport crude oil and refined petroleum products across Canada, from the producing provinces of Alberta and Saskatchewan to the consuming provinces and as far east as the northeastern United States. Crude oil is the most viscous liquid of those transported. All liquids require regular boosts in pressure to move them along.

Enbridge supplied the following information. Pumping stations are typically 40 miles (64 km) apart and each can boost the pressure from an intake value of 100 pounds per square inch (6.9×10^5 Pa) to an output value of 1200 psi (8.27×10^6 Pa). The difference between output pressure of one station and intake pressure at the next station pushes the oil along at a constant speed. A representative value of the viscosity of crude oil is 0.75 Pa·s. The diameter of the pipe along the main line is 48 inches (1.22 m). Two such pipes are often placed in parallel along the route where the demand requires it. Lower viscosity refined products are shipped in batches through the same pipes.

Using the above data:

a. Calculate the flow, in m^3/s, of crude oil through a single pipe 1.22 m in diameter based on a pressure difference of 15×10^5 Pa between stations. Since the oil industry publishes all its data in barrels/day (bbl/d), convert the flow from m^3/s to bbl/d. One barrel equals 158.98284 liters.

b. Calculate the speed of the oil in the pipe in m/s. How long will it take for a given barrel of oil to get from the Alberta border to a destination in Ontario 3500 km away?

The understanding of the physics of the oil pipeline is incomplete without calculating the energy required to transport this vital fluid across the continent. Begin by concentrating on a segment between two pumping stations.

c. What is the force applied to the oil in the pipe just after a boost by the pumps, and what is the force in the pipe just before the next boost? The difference between these forces drives the flow.

d. The energy needed to pump the liquid is the net force applied multiplied by the distance the liquid is moved. What then is the energy required to replace the total amount of oil between two stations? For this calculation use the distance the final drops of oil have to move from the highest pressure inlet side of the pipe to where these drops just reach the next pumping station.

e. How much oil, in m^3, is there in the segment of pipe between two stations?

f. How much work is done to move one liter of oil between two stations? How much energy is then expended to move one liter of oil over 3500 km? The pumps along the pipeline will not be 100% efficient; therefore your answer will underestimate transportation energy costs by 30% or more. This bit of reality is to be ignored for the rest of the problem. The energy content of one liter of oil is 3.8×10^7 J. What fraction of that energy is used for its own pipeline transportation?

g. Calculate the power (energy per second) required by the pumps to drive the oil from station to station. Engineers still like to express power in units called horsepower (hp), where 1 hp is defined as 750 J/s. What therefore is the horsepower requirement at each pumping station along the pipeline?

h. The pipes are tested for a maximum pressure of 1800 psi before they are put into service, but they are never used at this pressure. Suppose for a moment that Enbridge wanted to double the speed of travel of the oil. What pressure difference between stations would be required? What power would be required at the pumping stations?

NOTES

4 HEAT AND THERMODYNAMICS

4.1 TEMPERATURE AND ENERGY

4.1.1 Hot air balloons with passengers are now a familiar sight. They are not a useful mode of transportation because they can only move with the surrounding air. The principle (first applied for human flight in 1783) is to have a sufficiently large air-filled balloon, open at the bottom, with the means of heating the air inside the balloon to lower its density as compared to the surrounding air. A large enough volume with low density will float in a medium of higher density (cooler air). A sufficiently large volume of low density air in the balloon will allow a load to be lifted—the skin of the balloon and the basket with payload, burner, and fuel, hanging below. The earliest balloons had a continuous fire going and were rarely made of fire retardant materials. Modern balloons carry tanks of propane and a heavy-duty burner which forcibly directs a flame into the balloon above.

Recently toy hot air balloons have been advertised. These make use of solar energy instead of fuel to create the lift. Such a toy balloon has to be large and made from light-weight materials to be able to lift its relatively dense skin. The calculations are based on a hypothetical balloon with a radius of 51 cm. It has a plastic skin 6.3×10^{-3} mm thick with a specific gravity of 0.94, and a specific heat of 2.3×10^3 J/kg. The skin is black to absorb the solar energy which in turn heats the air inside the balloon. A small opening at the bottom of the balloon prevents the pressure from building up inside the balloon as it heats up. Through some miracle of engineering the balloon keeps its large spherical shape at all times without a frame.

The density of air at ambient pressure (sea level) and ambient temperature (20°C) is 1.30 kg/m^3. The specific heat of air at constant volume is 700 J/kg°C.

a. Calculate the volume of the balloon, the mass of air inside, and the mass of the skin of the balloon.

b. Bright sunlight shines on the balloon and somehow uniformly heats the air inside the balloon. As this air heats, it expands and some of it leaks out of the balloon keeping the pressure inside the balloon constant. Calculate the mass of the balloon and its contents when the temperature of the air inside the balloon reaches 35°C. Determine the buoyant force on the balloon. Is it enough for the balloon to rise or even to carry an extra load?

c. The skin of the balloon is black to absorb all incident light and heat from the solar radiation. On a bright sunny day as much as 1.3 kJ/s of solar radiation is directed perpendicular to 1.0 m^2 of surface. Calculate the area of the balloon that intercepts the sunlight? How long will it take for the skin of the balloon to accumulate enough energy from the sun to heat the air the required 15°? It will, of course, take longer than this for the air inside the balloon to heat uniformly.

4.1.2 On a bright, sunny day 1100 W/m^2 of solar radiation enters through the south-facing windows of a greenhouse. As long as the temperature difference between the interior of the greenhouse and the outdoors is not large, one can neglect the heat leaking out of the greenhouse and calculate the temperature increase based on the following data.

The south-facing window area receiving the light is 4.8 m wide by 2.5 m high and slanted to receive the energy from the sun at normal incidence. The floor of the greenhouse is 3.0 m by 4.8 m and the flat roof is 1.5 m by 4.8 m. Against the back wall stands a long container that is 0.49 m wide, 4.8 m long, and 1.2 m high. This container can be filled with water.

The specific heat of air is 700 J/kg K; the specific heat of water is 4.18×10^3 J/kg K. The density of the air is to be taken as 1.3 kg/m^3.

a. Assume the greenhouse is strictly empty space. Calculate the heat capacity of the air in the greenhouse.

b. Assume the container described in the introduction is filled with water. Calculate the heat capacity of the greenhouse.

c. With sunlight entering the greenhouse at the rate stated in the introduction, calculate the time it should take for the greenhouse to warm up by 2.5°C with and without the water in the container. Assume that no heat leaks out and that the temperature in the greenhouse is uniform at all times.

4.2 HEAT CONDUCTIVITY

4.2.1 In the non-metric language home insulation contractors use, a wall with a 10 cm thick fiberglass bat insulation has an *R* value (resistance to heat flow) of 16. A wall with twice that thickness of insulation will be R32. Currently recommended minimum insulation levels in most of Canada is R23 for walls and R36 for ceilings. The specifications for the same type of fiberglass insulation can be translated into metric units: the fiberglass insulation conducts heat at a rate of 0.040 W/m^2 of wall surface for a thickness of 1.0 m for each 1.0°C temperature difference between the two sides of the wall.

In Figure 4.1 the only outside wall of a room faces north and is 3.6 m wide and 2.5 m high. There are no windows. The insulation is to be considered entirely due to the 10 cm of fiberglass as described above.

a. Calculate the heat loss in joules through this north wall over a period of 24 h if the inside temperature is +20°C and the outside temperature is -20°C.

b. A high quality, double-glazed window is installed in the wall. The window has, again in non-metric language, an *R* value of 4.0. In metric units the specification for the window is a heat loss of 1.6 W/m^2 of window area per 1.0°C temperature difference. The window is 90 cm high and 120 cm wide. Calculate the heat loss of the room for the same temperature difference as in *part a.*

Figure 4.1

c. If the wall faced south instead of north there would be a gain from solar energy even in far northern locations. Suppose an average of 0.70 kW/m^2 of solar radiation is incident on the window for only 4 h per day, 72% of which enters the room as heat. Calculate the amount of heat entering the room by radiation over a 24 h period. Over a 24 h period, is the window an advantage from an energy conservation point of view? Would the window be an advantage on the north side of the house?

4.2.2 It is a standard engineering/physics problem to calculate heat loss from inanimate objects. This can be done with a high degree of accuracy., However, heat loss from the human body is more tricky but interesting. It is tricky because no two humans are exactly the same and, equally important, humans, like any other biological entity, have control mechanisms that change the response of the body to help cope with changes in the environment. The problem that follows explores the basic facts of heat loss and will make it obvious why and where control must come in. In other words, it is oversimplified to make the problem tractable with numerical assumptions that may be true for one person, but too large or too small for another.

The human body is of complex shape and varies with age and gender. For the rough calculations to follow, suppose that Barrelchest in Figure 4.2 has a body in the shape of a uniform cylinder, 1.6 m tall, diameter 0.15 m. An adequate diet for a physically inactive person is considered to be about 2 000 Calories per day (1 Cal = 4.184×10^3 J). Some of the energy from the food intake can be utilized to do mechanical work (See Problem 1.4.2), but for the couch potato under consideration assume all the food energy eventually gets dissipated as heat from the skin.

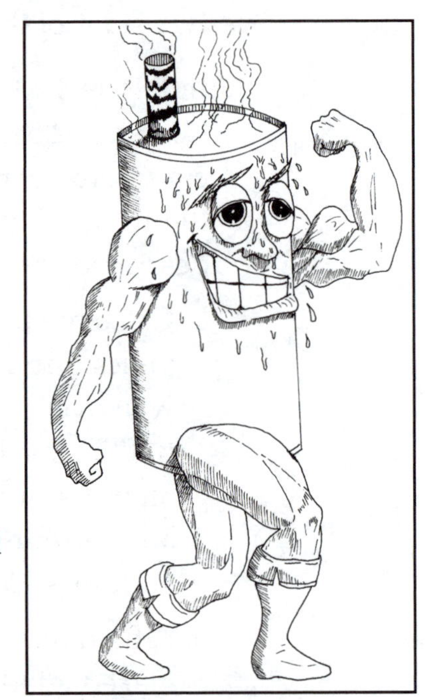

Figure 4.2

a. Determine, in watts, the required rate of heat loss for Barrelchest to stay in temperature equilibrium. Based on his assumed body size and shape, calculate the required heat loss per m^2 of body surface to keep his body temperature constant.

b. As the ambient temperature decreases people wear more clothing, when the temperature rises, clothing is shed as much as modesty and the law will permit. It is possible to estimate the average insulating quality of the human skin and the immediately adjacent layer of air. The interior of the body is kept at a uniform 37.0°C. The temperature at which an unclad body at rest can feel comfortable in the shade is about 28°C. (This temperature is highly dependent on subcutaneous fat, hair, and other personal matters.) Based on Newton's law of cooling, $\frac{\Delta Q}{\Delta t} = \frac{K}{L} A \, \Delta T$ where ΔQ is the energy conducted from the warm to the cold reservoir in time Δt, A is the area through which the heat energy flows and ΔT is the temperature difference that drives the heat flow, determine the ratio K/L and its reciprocal, L/K. The ratio K/L is the combined effect of heat conductivity K and thickness of the layer, L. Neither K nor L can be determined by themselves, but their ratio is obtainable from measurement, given the temperature difference and magnitude of required heat flow. The ratio L/K is known as the thermal resistance.

c. Barrelchest moves to an air-conditioned room at 20°C. What thermal resistance does he require to feel comfortable? Remember that his rate of metabolism is to remain constant.

d. Slowly the situation is complicated to make it more realistic. A certain portion of the body remains uncovered unless the temperatures are extreme or religious customs prevail. Face and hands tend to be bare. As a rough estimate, each hand with the fingers held tight against each other, exposes an area of 10 by 20 cm on each side. The face exposes a circular area of radius 16 cm. Given this guesstimate and the thermal resistance of the skin as estimated in *part b*, calculate the heat loss of the exposed area at 0°C. How well insulated must the rest of the body be to keep Barrelchest's the heat loss to 2 000 Cal per day? Calculate the required thermal resistance of the clothing.

   NOTE: One layer of ordinary clothing material has a thermal resistance of approximately 0.1 m^2 C°/W. The insulation value of clothing lies mostly in the stagnant air trapped between the fibers of the clothing and between layers of clothing. Quilted fabric filled with down is more effective than woven fabric because more air can be trapped for the given weight of a garment. A down jacket or sleeping bag has a thermal resistance of approximately 1 m^2 C°/W. The clothing trade even has its own unit of thermal resistance called the tog (one tog = 0.10 m^2 C°/W). Thus an ordinary layer of clothing has a thermal resistance of about 1 tog, while a down jacket has a thermal resistance of close to 10 togs.

e. When the body cannot lose sufficient heat by conduction, then evaporation (sweating) kicks in. This has the unfortunate effect of causing the clothing to become wet, which in turn reduces the thermal resistance of the clothing. Try now to estimate the amount of clothing (i.e. thermal resistance required) a cross country ski racer should wear during a race at –10°C. The athlete's body burns food energy at 1.5 kW, of which 0.3 kW is turned into mechanical energy of motion, the rest is heat that must be dissipated. Although cross country ski racers, like long distance runners, are lean and slender, stick to the dimensions of Barrelchest in this problem, and ignore the effect of hands and face.

Machines, electronics, and biological systems must be kept at an even temperature to operate reliably. Automobiles have radiators with fans, electronic equipment can often make do with natural ventilation, and humans sweat. Heat transfer systems are of great technological importance and come in different forms. Copper tubing is used to transfer heat from fluids because it is relatively easy to use and readily available. Silver has even better heat transmission properties but is rarely used because it is too expensive. Diamonds have even higher heat conductivities than the best metals. They are very expensive, but they are widely used in cooling systems for miniature electronic applications. Boiling water keeps foods at an even 100°C on the cooking stove.

4.2.3 An open metal (aluminum or copper) pot sits on a gas stove. The diameter of the pot is 16 cm, there are 11 cm of water in the pot at the start, and there is a hot flame underneath. When the flame is small the water barely bubbles as it boils. With the flame set to its maximum value the water bubbles vigorously.

a. The water in the pot is boiling to prepare a pot of tea, but the conversation which was to accompany the tea becomes so interesting that the pot and stove are completely forgotten. The flame transfers 1200 W into the pot. How long does it take to bring the water to a boil from 15°C? How much longer does it take for the open pot to boil dry? Ignore all heat losses from the pot other than the evaporation of steam at 100°C.

b. Heat enters the pot from the bottom at a rate of 1200 W. It is rapidly transferred in the form of bubbles of steam through the 11 cm thick layer of water. If the heat flow through the water was like the heat flow through a solid like copper, what would be the heat conductivity in J/s m C° of the solid, 11 cm thick and 16 cm diameter?

HINT: The cross-sectional area and distance over which the heat travels are easy to establish, but is there a temperature difference between the top and bottom of the water in the pot? For the purpose of the problem, suppose quite arbitrarily that the top of the boiling water is 0.1 degree cooler than the bottom.

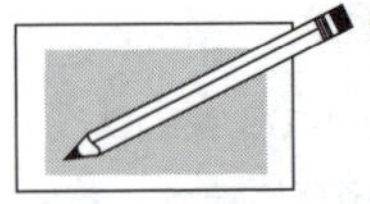

c. An engineer from an "intellectually challenged" part of the globe has come to the conclusion that the escaping steam from the standard stove wastes too much energy in bubbling through the boiling water.

The revised design is a burner above the pot with suitable reflectors to reflect all the heat from the burner to the water surface below. Discuss the advantages vs. disadvantages of the proposed design.

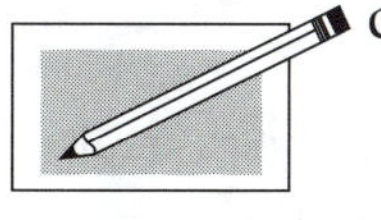

d. The weightlessness in space stations presents difficulties in cooking meals. Explain why this is so.

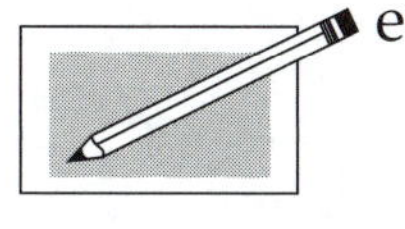

e. Water is easy to bring to a boil over an open flame. A thick pea soup or goulash is difficult to heat evenly. Why?

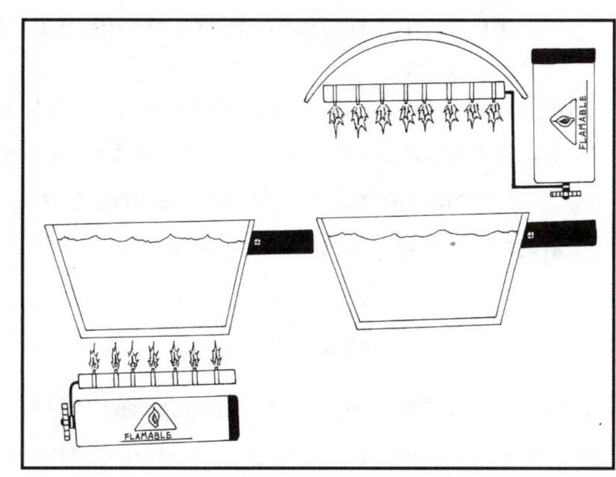

Figure 4.3

There are differences between boiling water in an open pot and in a covered pot. In addition, there are differences between a covered pot and in a tightly sealed pot of boiling water. The contents of the covered pot remain at atmospheric pressure, while the pressure in the sealed pot increases as water is converted into steam. The sealed pot becomes the pressure cooker.

4.2.4 Maple syrup is made by boiling the sap of maple trees in large open kettles. One way of preparing rice is by cooking it in boiling water in a covered pot.

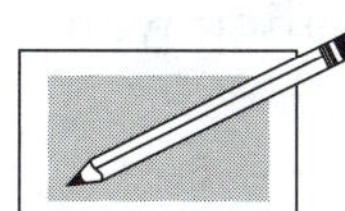

a. Explain the difference in treatment.

b. The observant kitchen scientist will notice that it takes less heat to keep the contents of a covered pot boiling as it takes to keep the same contents boiling without a cover. Explain the why this is so.

Mountain climbers are aware that meals take longer to cook as they climb higher and that coffee and tea do not seem to be as hot as they are at sea level. The reason is that atmospheric pressure depends on altitude (see Section 3.2) and this in turn influences the boiling point of water, or any other liquid. For the cook there are two considerations. Many foods, such as stew meats, need tenderizing requiring extended periods of cooking. This process is speeded up by higher boiling temperatures. For other foods, the main requirement is that the temperature on the inside of the food reaches a temperature high enough for a chemical reaction to take place. The boiled egg is a good example. Both types of processes are aided by higher boiling points. The egg needs less time to cook because the higher water temperature speeds the heat flow into the interior of the egg. The stew becomes tender more rapidly because the fibers soften more rapidly at the higher temperature. The pressure cooker is a device that makes use of the generated steam to build up the pressure in the vessel and therefore increase the temperature of the liquid and other contents. The design of pressure cookers includes safety valves that limit the pressure and prevent explosions.

The reduction of the boiling point also has practical applications. Some food must be desiccated (dried) using low heat in order to preserve vitamins. Such food is placed in an airtight vessel, the pressure is reduced with vacuum pumps, and the water inside the food boils off at safe temperatures. Freeze drying is another form of this technology.

The variation of boiling point with pressure can be derived from first principles and is dealt with in physical chemistry. Here the interest is in observable kitchen physics and the relationship between pressure and the boiling point of water is taken as an empirical (experimentally found) relationship:

$$\log_{10}(P) = \frac{\phi \Delta T}{373.1 + 1.15 \Delta T},$$

where P is the pressure measured in atmospheres, ϕ is an empirical constant equal to 5.70 for water, and ΔT is the increase in the temperature of the boiling point of water. The number 373.1 that appears in the equation is the boiling point of water under standard atmospheric pressure in °K.

4.2.5 A certain biological specimen must be desiccated for long-term storage, however an important enzyme will be destroyed by any temperature above 40°C. At what pressure will the boiling point of water remain at 36°C or less?

4.2.6 A pressure cooker for kitchen use is designed to maintain an internal pressure of 2.0 atmospheres. Calculate the temperature of boiling water under that pressure.

4.3 IDEAL GAS LAWS AND KINETIC THEORY

4.3.1 The title of the book *Fahrenheit 451* by Ray Bradbury is the temperature at which paper ignites spontaneously. *Fahrenheit 451* deals with book burning as one symptom of an oppressive government. In this problem you are asked to explore the physics of an unusual way to start a fire.

A small piece of paper is placed in a cylinder sealed by a piston. A blow with a heavy mallet drives the piston down in the cylinder to compress the air surrounding the paper. Figure 4.4 shows the arrangement. Because the piston is driven down rapidly, a negligible amount of heat from the rise in temperature of the gas will escape during the compression phase. The process can be considered adiabatic.

The diesel engine makes use of this process in real life.

a. Convert 451°F to the Kelvin scale.

b. The gas and the paper are initially at 20°C and at atmospheric pressure. By what factor must the volume of the air in the cylinder be reduced for the paper to reach the ignition temperature?

c. Calculate the air pressure in the cylinder at the ignition temperature.

Figure 4.4

4.3.2 The derivation of the ideal gas law, even based on the simplest statistical mechanics model, predicts a specific relationship between the average kinetic energy of the atoms (or molecules) of a gas and the absolute temperature. It is only a small step to make the additional assumption that all the atoms or molecules in the gas have the same mass and to obtain the average speed of the molecules as a function of temperature. The theories of sound are based on compressional waves in gases which in turn means that one layer of gas is able to push on the next layer, bounce back, and receive the next vibration to transmit. It is difficult to perceive a theory of sound where the speed of the acoustic signal exceeds the speed of individual particles in the gas. A detailed statistical model of the ideal gas law predicts that the velocity of sound must be proportional to the average speed (more correctly, v_{rms} the root mean square velocity) of the molecules. The predicted factor of proportionality is $\gamma^{1/2} = (c_p/c_v)^{1/2}$ where c_p and c_v are the specific heats of the gas at constant pressure and constant volume respectively. Specifically the theory predicts $v_{sound} = (\gamma/3)^{1/2} v_{rms}$.

a. From Avogadro's number and the atomic or molecular weights of hydrogen, helium, and nitrogen calculate the average speeds (v_{rms}) of the atoms or molecules for the three gases at -40°C and at +20°C.

b. Predict the speed of sound in the three gases at the two temperatures. Compare your results with published values and with the accepted temperature dependence of the speed of sound.

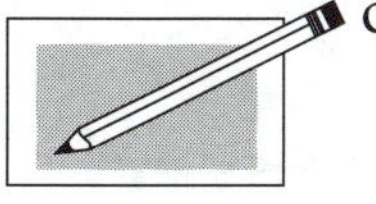

c. The results obtained from *parts a* and *b* are independent of the pressure in the gas. The introductory paragraph explains that the transmission of sound must move from gas layer to gas layer and therefore involves collisions of molecules in one layer with molecules of the next layer. Explain how collisions between molecules should depend on gas pressure and what frequencies (or wavelengths) in the transmission of sound will be affected first as the pressure drops. The concept involved is called the mean free path of gas molecules, the average distance a molecule travels in a gas between successive collisions.

4.4 HEAT ENGINES

4.4.1 Nuclear reactions in a nuclear power plant take place at temperatures of millions of degrees on the atomic scale. The temperature from one hot reacting atom is averaged out to many neighboring atoms kept cool by the steady circulation of cooling fluids. The heat absorbed by the cooling fluids is used to drive the turbines which produce electrical power.

The majority of electrical power produced in Ontario comes from nuclear power plants. Deuterium oxide (D_2O), also known as heavy water, is the coolant in these CANDU reactors. The maximum temperature at which heat can be extracted from the heavy water coolant in a reactor is 290°C. In an experimental reactor at the Whiteshell Nuclear Research Establishment in Manitoba, a wax-like organic coolant has been substituted for the heavy water. It allows the reactor to operate at 400°C instead of 290°C.

Assume that the real heat engine used to extract electrical power from the reactor operates at 75% of the efficiency of an ideal Carnot engine and that the cold reservoir is at 40°C.

a. Calculate the ideal and real thermodynamic efficiencies of the heavy water nuclear power plant.

b. A real nuclear power generating plant consists of four or more individual units each of which produces heat from nuclear reactions at a rate of 2.3×10^9 W. Calculate the electrical power each unit generates and how much waste heat is ejected into the environment, in this case into Lake Ontario.

c. Calculate the ideal and real thermodynamic efficiencies of the reactor if the organic coolant is substituted for the heavy water. What percentage improvement can be expected?

4.4.2 Air conditioners are practical forms of heat engines that use electrical energy to remove heat from a cold reservoir, the room, and deposit heat into a hot reservoir, the outdoors. The ideal air conditioner would have the efficiency of a Carnot engine. In practice, an appreciable fraction of the energy supplied to drive the device is wasted through friction and the leakage of heat through the machine from the hot to the cold reservoir.

Air conditioners are also designed to remove moisture from the air in the room. This aspect requires additional energy and will be ignored in this problem. Like the Carnot engine, the air conditioner is reversible—it can remove heat from the already cool outside air and deposit this energy plus the work done in driving the device into the room. The air conditioner in this mode is an efficient space heater as long as it is not too cold outside.

In the calculations to follow, first assume that the air conditioner is a perfect heat engine with the efficiency of the Carnot engine. At the end of the calculations reality should be brought in by increasing the electrical energy required to run the air conditioner by 30% over the estimates made on the basis of Carnot efficiency. In the comparison between electrical heating and using a heat engine, keep in mind the capital cost of air conditioners vs. electrical radiators. More important, keep in mind the gains expected from improved home insulation.

The numbers to be dealt with are based on a poorly insulated house in a warm region of the world where frost is a rare occurrence. The house is essentially a rectangular box with a 10 m by 20 m floor area and 4.5 m high walls. Ceiling and walls have insulation equivalent to a thickness of 2.5 cm of fiberglass with a heat conductivity k of 0.040 J/s m C°. (In home insulation jargon these are R4 walls and ceiling.) Ignore the heat gain or loss through the floor.

a. On a hot summer day the outside air temperature is 40°C. Inside the house the temperature is maintained at 20°C. Calculate the rate at which heat leaks into the house through walls and ceiling. On a cool winter day the outside air temperature is 0°C but the temperature inside is still kept at 20°C. At what rate does heat leak from the inside to the outside?

b. The house is heated on that cool winter day by electric heaters. How many kWh are required to keep the house at 20°C over a 24 h period? At a cost of $0.083/kWh for the electricity, calculate the cost of heating the house for the day.

c. The perfect air conditioner is run as a heat engine to remove heat energy from the outdoors and to deposit this energy in the house. The work done to operate the ideal heat engine is also deposited in the house. How much work must be done to supply the inside of the house with enough energy to keep it at 20°C for 24 h? The air conditioner is not, however, an ideal heat engine and therefore 30% of the energy input may be wasted. What will the electrical cost be to keep the house heated for 24 h with this more realistic device?

d. The air conditioner is now run for the purpose it was designed, to cool the room. Again assuming the air conditioner is a perfect heat engine, how much work must be done over a 24 h period to keep the room cool on a hot (40°C) summer day? Again assuming only 70% of the energy you pay for cools the house, how much will it cost to run the air conditioner for 24 h?

4.4.3 Cogeneration is an increasingly popular option for electric power plants. The principle of cogeneration is to make the best possible use of the fuel consumed by balancing the sales of electric power output of the power station with sales of usable heat for industrial or domestic use. For example, a car wash needs large quantities of warm water and electricity for day-to-day operations. A carefully designed power plant uses the energy content of fossil fuels to create work in the form of electricity. However, according to the laws of thermodynamics, a large fraction of the fuel's energy is rejected as waste heat. This waste heat could be used to heat the cold water in the car wash operation. Power companies that sell only electricity are interested in maximizing electrical power output, therefore these companies use the lowest possible cold reservoir temperature. Cogeneration power plants deliberately use a higher temperature cold reservoir and distribute the remaining heat as heat and hot water in nearby homes or industry for profit. Electrical energy can be sold at approximately five times the price of the same amount of energy in the form of heat.

a. A certain power plant operates as if it were an ideal (Carnot) engine with a hot reservoir of 330°C. The heat input is at a rate of 20 MW. Thanks to the cooling capacity of a nearby river, the cold reservoir can be kept to an average of 30°C. Calculate the electrical energy produced over a period of one week. If the electrical energy is sold at a rate of $ 0.082 per kWh, then how much money comes in per week from electricity sales?

b. A survey has found that a market exists for all the rejected heat from the power plant as long as the heat (hot water) can be delivered to consumers at 50°C or higher. The consequence for the power plant is that the cold reservoir must have a temperature of 60°C instead of the previous 30°C. How much electrical energy can the plant produce per week with the warmer cold reservoir? How much heat can be generated for distribution? If this heat energy is sold at 1/5 the price of electricity, then what will be the weekly income from the total output of the plant?

c. If the total heat production is converted strictly into heat for domestic heating or industrial use, as is the case in all home furnaces and hot water heaters, then the 200 MW heat input of the plant becomes 200 MW heat output. At 1/5 the price of electricity, what would be the weekly income of the plant?

NOTES

5 WAVE MOTION AND SOUND

5.1 FREQUENCY, WAVELENGTH, AND SPEED OF SOUND

The definition of sound should be those vibrations in the air that human ears can detect. The healthy and youthful human ear is sensitive only to the frequency range approximately between 20 Hz and 20 kHz; therefore those should be the limits to what we call sound waves. However, microphones and speakers can, depending on design, easily be used in a wider frequency range, and can also easily create and detect vibrations in solids and liquids. The principles of behavior of these other waves are the same. Of particular interest are "sound" waves with a frequency of 10 or more times what the human ear can detect. These waves are called "ultrasound" and are important to bats and in medical applications. Some of the applications of these vibrations are included in the problems of this section.

5.1.1 The speed of sound in air is approximately 345 m/s while the speed of light is approximately 3.0×10^8 m/s. During a thunderstorm the crack of thunder and the bolt of lightning are created at the same place and at the same time. It is possible to determine how far away that event took place by measuring the time it takes for the thunder to be heard after the flash of lightning has been seen.

a. How far away was the lightning strike if the thunder is heard precisely 3.26 s after the flash is seen? What is the distance if the time interval is approximately 5 s?

b. Suppose the speed of light were less than 3.0×10^8 m/s. How "slow" would the speed of light have to be for an observer with a stopwatch to be in error by 1% in judging a distance of 1.3 km using the technique in *part a*?

5.1.2 The speed of sound varies with altitude and temperature in predictable ways. Aircraft response changes with the speed of sound; therefore pilots of jet aircraft prefer to specify the speed of their airplanes in relation to the speed of sound rather than as m/s or km/h. The speed unit used is the Mach number. Mach 1 is the speed of sound, Mach 3 is 3 times the speed of sound, and Mach 0.75 is 3/4 of the speed of sound. Warp speeds are inventions from science fiction.

a. Commercial jets fly at about Mach 0.75. Convert that speed into m/s and km/h based on the speed of sound being 330 m/s.

b. The speed of sound in air v as a function of temperature can be expressed as

$$v = 331\sqrt{\frac{T}{273.15}}$$

where T is the temperature on the Kelvin scale. The surface temperature at the airport in Edmonton, Alberta, can vary from +40°C in the summer to –40°C in the winter. Similar temperature changes occur as an aircraft climbs from the airport to its cruising altitude of 10 000 m. Calculate how much Mach 1 varies over this temperature range.

The equation $v = \lambda f$ states the relationship between the speed of the wave v, the wavelength λ, and the frequency f. The mathematical interpretation is that, given any two of the variables in the equation, the third variable can be determined. In the physical world the priorities are different. Once the material is specified, the velocity of the wave is set. For example, the speed with which a wave travels along a taut string can be calculated from the mass per unit length of the string and the tension in the string. A wave is created by shaking a portion of some material at a frequency of choice. This procedure creates the wave that begins to travel through the material. The wavelength that results can be predicted from the properties of the material and the frequency that has been applied.

Wherever there is an interface, i.e. where two different materials meet, the frequency of the disturbance crossing the interface remains the same. The second material is shaken at the same frequency as the first one is shaking. However, if there is a change in wave speed because of the change in material, the wavelength has to change to keep $v = \lambda f$.

Two additional factors come into play. The expression

$$v = \sqrt{\frac{T}{\mu}}$$

for the speed of a wave on a string v as a function of tension T and mass per unit length μ, implies that the speed with which a disturbance travels through a given material is independent of frequency. The equation is based on approximations made in the derivations. Experiments with light and sound show otherwise. For example, the speed of light in glass changes noticeably over the range from violet to red and is responsible for the spectrum observed with prisms. The second factor of interest for this section is that a change in speed of a wave at an interface causes a reflection of part of the energy of the wave. The strength of the reflection increases with increasing change in wave speed over the interface.

Our knowledge of the structure of the crust of the earth, many kilometers below the surface of the Earth, is obtained from seismic studies. One approach used is to set up delicate sensors (seismometers) in widely separated laboratories around the world. The researcher waits for an earthquake to take place somewhere in the world and carefully analyzes the amplitudes and arrival times of the waves that may have travelled through some structures of interest in the Earth's interior.

A second approach is more practical for short distances. A series of vibration detectors (a specialized form of earphones) is laid on the ground over a distance of up to one kilometer. A small explosion is set off, and the vibrations detected in each individual detector are measured and recorded. The detectors first pick up the strong signal that goes directly from the explosion site to the earphones, but that signal bears little information. The signals received later, which are due to reflections from underground structures hundreds or thousands of meters below the surface, are analyzed in great detail. Oil and gas reservoirs are inferred from studies of this type.

5.1.3 There are so many possible underground structures that it is difficult to infer the shape of the structures from the seismic data. On the other hand it is not difficult to predict what the record should look like for specific assumptions about the underground structures. The following oversimplified situation illustrates a few of the principles involved.

A seismic array consists of five detectors in a straight north-south line, 200 m apart. The ground is perfectly flat and the speed of sound everywhere in the ground is a uniform 6.10 km/s. A flat discontinuity exists at a uniform depth of 1.20 km below the surface causing a partial reflection of the energy of any disturbance created above it. The seismic crew sets off a small explosion near the surface 200 m north of the line of detectors. The arrival times and magnitudes of signals due to the disturbance are recorded from each of the detectors. Figure 5.1 shows part of the scene.

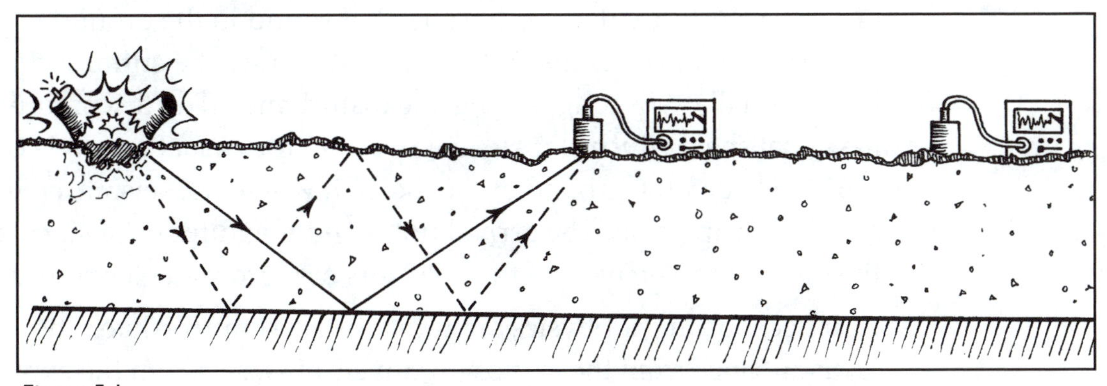

Figure 5.1

a. Reflections in geophysics are rarely (if ever) mirror-like. The disturbance from the explosion travels outward from the explosion site in all directions and hits the discontinuity. From there reflections will occur in all possible directions, including upward to the detectors. Sketch a diagram showing the explosion site, one of the detectors, and some of the paths the disturbance can take from the explosion site to this detector. Label the shortest path(s).

b. Calculate the first expected arrival times of the direct and the reflected signals at each of the five detectors.

c. Assume only 2.0% of the energy is reflected at the interface, regardless of direction of incidence, and that the ubiquitous $1/r^2$ law operates. Estimate the relative energies of the signal expected at each detector.

d. Even for this simple geometry more signals are expected. Figure 5.1 shows that there will be multiple reflections to each detector. Calculate arrival times and expected energies at the third detector.

e. Sketch a graph of the signal vs time as it might be recorded at the third detector. Consider the explosion as a pulse of negligibly short duration. How will the duration of the direct and reflected pulses compare to each other?

5.1.4 There are so many possible underground structures that it is difficult to infer the shape of the structures from the seismic data. On the other hand it is not difficult to predict what the record should look like for specific assumptions about the underground structures. The following oversimplified situation illustrates a few of the principles involved.

NOTE: Though this problem seems to repeat Problem 5.1.3, this problem deals with a slightly more complicated underground structure.

A seismic array consists of five detectors in a straight east-west line, 180 m apart. An explosion will be set off 180 m to the west of the array. The ground is perfectly flat and the speed of sound in the ground is a uniform 6.20 km/s. A discontinuity exists under the entire section of land, partially reflecting the seismic disturbance. The discontinuity is like a flat sheet which slopes downward to the east at an angle of 25°. It is 670 m below the ground at the location where the seismic crew will set off the explosion. The arrival times and magnitudes of signals due to the disturbance on each of the detectors are recorded electronically.

a. Reflections in geophysics are rarely (if ever) mirror-like. The disturbance from the explosion travels outward from the explosion site in all directions and hits the discontinuity. From there reflections will occur in all possible directions, including upward to the detectors. Sketch a diagram showing the explosion site, one of the detectors, and some of the paths the disturbance can take from the explosion site to this detector. The shortest path from source to detector, after reflection from the discontinuity, is identical to the path that would be followed by a light beam from a source to a detector after hitting a plane mirror. Identify such a path in the diagram.

b. Calculate the first expected arrival times of the direct and the reflected signals at each of the five detectors.

c. Assume only 2.0% of the energy is reflected at the interface, regardless of direction of incidence, and that the ubiquitous $1/r^2$ law operates. Estimate the relative energies of the signal expected at each detector.

d. Even for this simple geometry more signals are expected. Figure 5.2 shows multiple reflections to each detector. Calculate arrival times and their expected energies at the fourth detector.

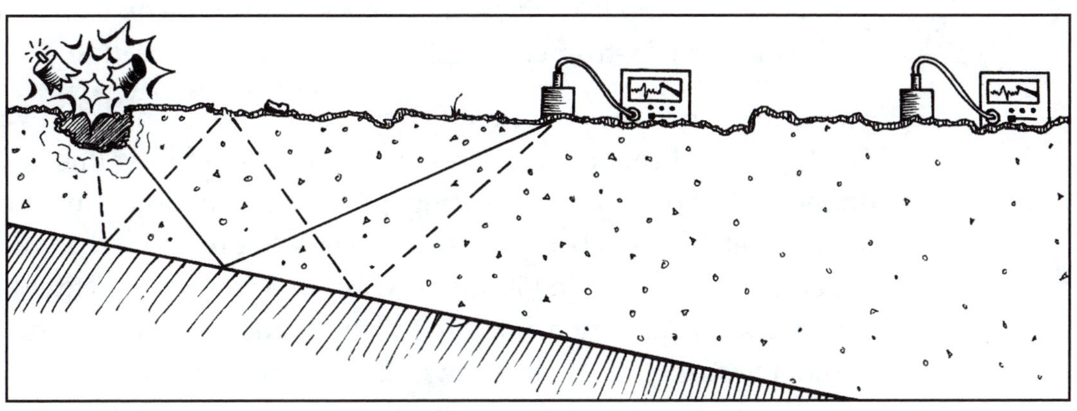

Figure 5.2

e. Sketch a graph of the signal vs time as it might be recorded at the fourth detector. Consider the explosion as a pulse of negligibly short duration. How will the duration of the direct and reflected pulses compare to each other?

5.1.5 The medical diagnostics procedure of ultrasound imaging uses pulses of high frequency (3.5 MHz or 5.0 MHz) vibrations to probe the body. The pulse is sent into the body from a hand-held device and is transmitted by the body tissue. Energy from the beam is reflected whenever there is a change in the type of tissue. In the simplest version of this machine the hand-held device sends out the pulses, picks up the reflections, and electronically registers the arrival time of the reflections from the body.

Modern medical ultrasound equipment can register the directions and distances from which the reflections come. The shapes of organs and any irregularities are displayed on a computer screen. To avoid overlap of the pulse going out with pulses coming back, the system is in the "send" mode for 1.0% of the time, while a pulse is created and sent into the body. For the rest of the time the system is in the "receive" or "listen" mode, registering the amplitudes of the reflected waves.

a. The speed of sound in human tissue is approximately 1500 m/s. The diagnostician may want to observe to a depth of 25 cm in a body. Calculate the length of time the instrument must be in the receive mode in order to make the required observations.

b. A separate requirement may be to look at structures of 2 mm in thickness. Calculate the maximum time the device can be in the send mode in order to detect such small details. How many wavelengths at 3.5 MHz does this represent?

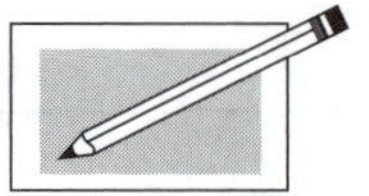

c. Before taking the measurements, the person doing the ultrasound scan always applies a thick layer of a specific type of gel between the body and the instrument. Explain why this is required.

5.1.6 The string telephone is a scientific toy. The device is shown in Figure 5.3. The tops have been removed from two tin cans to form either the microphone or the speaker. A thin string, preferably a monofilament fishing line, is stretched taut between the two cans. Sound from the voice sets the bottom of one can in vibration. This vibration is transmitted to the attached string and then along the string to the second can. The vibration of the string sets the bottom of the second can into motion, which in turn moves the air, and these vibrations are then audible to the listener.

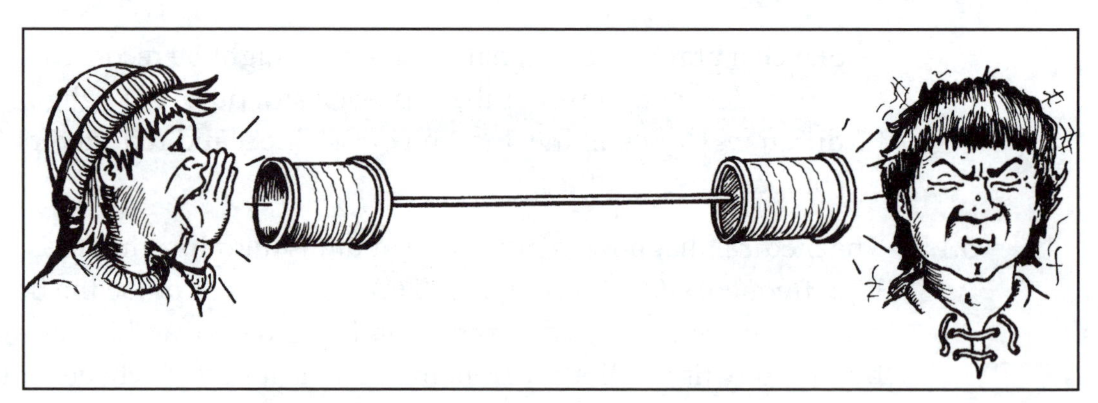

Figure 5.3

The specifications for a suitable monofilament fishing line are a mass of 14.7 g per 100 m length and a maximum load-bearing strength of 9.20 N.

a. A toddler screams into the first can at a steady pure note of C at two octaves above middle C (which corresponds to a frequency of 1046 Hz). How fast will this sound move as a wave in air? What is the wavelength of that note in air?

b. Calculate the frequency, wavelength, and speed of the note as transmitted along the string with the string held at its maximum rated tension.

c. Calculate the frequency at which the bottom of the second can will vibrate, and the frequency, wavelength, and speed of the sound that the observer hears over this telephone.

5.1.7 Compact disc players use tiny lasers to "read" the bumps on compact discs. The wavelength emitted by one of these lasers is 781.35 nm in air. The laser beam is directed to the disc through glass lenses, and the light is created in a crystal of gallium arsenide. Instead of listing the speed of light in different materials, it has become the rule to list the ratio of the speed of light in a vacuum to the speed of light in different materials. That ratio is called the index of refraction. The speed of light in a vacuum is exactly 299 792 458 m/s.

a. Calculate the wavelength of the laser light in a vacuum, in the glass lens, and in the gallium arsenide crystal. The index of refraction of air is $(1 + 2.748 \times 10^{-4})$, the index of refraction of this particular glass is 1.51106, and the index of refraction for gallium arsenide is 3.54. Use the correct number of significant figures in your answer.

b. Determine the frequency of the laser light in air, vacuum, glass, and gallium arsenide, paying particular attention to the correct number of significant figures.

5.2 DISSIPATION AND ATTENUATION OF SOUND

The cliché "time heals all wounds" may be translated as "distance attenuates sound." Two processes work together to make loud noises tolerable with increasing distance. The simplest to deal with is the $1/r^2$ law. That is, a point source of radiation of any kind will spread equally in all directions, always covering the surface of a surrounding sphere. A second process involved at the same time is dissipation. The energy of the source creates random movement in the surroundings which steadily decreases the energy remaining in the beam.

Sound may be dissipated by clothing, rugs, plants, acoustic tile, and the molecules of air. Fog dampens sound; dry air causes little dissipation. Sound carries a long distance over a still lake; it quickly dissipates in a forest. Dissipation is best described by the exponential relationship

$$I = I_o e^{-ar}$$

where I is the sound level in watts per square meter at the distance r from the source level I_o also stated in watts per square meter. The constant a quantifies the dissipation. A large value of a implies a strong decrease in the sound level over a short distance. An expression combining attenuation and dissipation can be written as

$$I = I_o \frac{e^{-ar}}{r^2}$$

5.2.1 A rock band plays at a sound level of 110 dB as heard by George 20 m away.

a. Calculate the source level I_o, assuming no dissipation.

b. Still assuming no dissipation, and therefore depending strictly on the $1/r^2$ law to decrease the sound level, how far from the band would Henry and Irma have to be in order to whisper to each other at a sound level of 30 dB?

c. The dissipation constant a is specified at 2.5×10^{-3} per meter. Assume this dissipation constant is the only reason the sound decreases with distance from the band.What is I_o under these circumstances? Calculate how far Henry and Irma have to be from the band in order to whisper to each other at a sound level of 30 dB.

d. Now assume that both dissipation e^{-ar} and the $1/r^2$ law are in effect. On this basis, recalculate the source level I_o required to hear the band at 110 dB sound level at the distance of 20 m.

e. **CHALLENGE:** Recalculate how far from the band Irma and Henry must be in order to be able to whisper to each other at the 30 dB sound level when both dissipation e^{-ar} and the $1/r^2$ law are in effect. Which plays a more significant role at this greater distance, the dissipation e^{-ar} or the $1/r^2$ law? Why?

5.2.2 The noise from a highway is a steady roar when there is continuous traffic. Cars and trucks create noise over the entire length of the road. When one car has passed a point, the second car is already in place. Mathematically such a highway is no longer a point source, but rather a line source. The mathematical consequence is that the noise level no longer drops at $1/r^2$ but instead at $1/r$.

Traffic experts have made numerous measurements and have come up with an equation for road noise which includes such factors as traffic density, proportion of trucks to cars, road surface, and average speed of the traffic. The base number calculated is the noise level at 30 m from the road followed by corrections for the distance of the observer from the road. The empirical equation used to obtain an estimate of the road noise L in dB at 30 m from the road is stated as

$$L = 25 \log_{10}(V) + 10 \log_{10}(N) - 26$$

where V is the speed of the traffic in km/h and N is the number of cars over a 24-hour period. A truck is counted as more than one car. Once the road noise at 30 m has been calculated, one subtracts

$$10 \log_{10}\left(\frac{D}{30}\right) \text{ in dB}$$

to obtain the road noise at the distance D from the road, also measured in meters. Shrubs and sound barriers require additional corrections which we ignore.

The traffic experts claim a noise level of 55 dB or less is satisfactory to most people.

a. Start with a calculation of the noise at 30 m away from a mildly busy road bordering a residential area. The traffic volume is 20 000 cars per day moving at a steady 50 km/h. Is that noise level satisfactory according to the experts?

b. As cities grow, traffic increases. Additional lanes are added and a higher level of traffic continues later into the night. Calculate the noise level 40 000 cars will produce at a steady speed of 50 km/h. At what distance should you be from that 40 000 car road to have the same noise level as calculated in *part a*?

c. A four-lane limited access highway has a larger carrying capacity and higher speeds. For a 24-hour period 100 000 cars moving at 90 km/h is a reasonable assumption on which to base further calculations. Calculate the distance required to have the same noise level as in *part a*.

d. The dB scale is useful because the sensitivity of the human ear follows a logarithmic law. It is also useful at times to keep track of the actual sound energy reaching an observer. Zero dB corresponds to an intensity of sound of 10^{-12} watts per square meter; 100 dB corresponds to 10^{-2} W/m^2. Refer to your textbooks for further details on dB. Show that the expressions stated in the introduction to this problem imply a doubling of the sound intensity due to a doubling in the number of cars for a given speed of the traffic. Show also that they imply that the sound intensity at the observer is inversely proportional to the distance, not to the square of the distance from the road.

e. The human ear begins to notice a difference in sound levels when there is an increase or decrease of 3 dB. Show that, regardless of starting level, a change of 3 dB implies an increase or a decrease by a factor of two of the intensity as measured in W/m^2.

5.3 WAVES OF UNSPECIFIED SHAPE

A detailed mathematical description of wave motion combines two principles. The first principle is that a disturbance created at one location will be noticed at a later time at an other location. The second principle is that the disturbance itself is precisely repetitive.

Here are some examples of the first principle. A boat crossing the middle of a lake will eventually cause ripples along the shore. An earthquake in California will register on a sensitive seismometer in Europe, but with a delay that depends on the distance from source to detector. A momentary disturbance whose shape is irrelevant is called a pulse. It may be a flash of light, a shout, an earthquake, or a rock falling in the water. It could also be a clothing style adopted in Paris or an influenza virus first identified in Hanoi. The disturbance can be propagated in vacuum, in air, in a solid, in a liquid, or in the human population. It will often change shape and strength, but it will be recognized as a disturbance away from the source after some elapsed time based on a characteristic speed of propagation.

This section is a collection of problems that deals with pulses of unspecified shape. In section 5.4 the second mathematical requirement of a true wave phenomenon will be added, that the disturbance itself is repetitive like simple harmonic motion and that its shape remains the same as it travels from source to detector. Ripples created in the middle of the lake are still ripples as they come toward shore.

5.3.1 Earthquakes occur when two land masses shift suddenly to relieve pressure that has built up over time. There is often local surface damage, but for the purposes of these problems consider the earthquake as a wonderful source of seismic pulses which permit the exploration of the earth's properties deeper than humankind can ever hope to dig or drill.

The originating pulse consists of both compressional and shear components, each of which can travel along the surface of the earth and through the crust of the earth. Seismometers far away detect the disturbance. The trained seismologist can distinguish between the different components and knows the speeds with which each of them travels. A calculation based on seismic data from a single seismic station can reveal the distance from the seismometer to the source. Three or more seismometer records from stations placed far enough apart can be used to pinpoint the location of the earthquake.

a. Compressional waves travel at a speed of 6.0 km/s through the earth's crust. Shear waves travel at a speed of 4.0 km/s in the same environment. Given that the arrival times of the two are 20 s apart, determine the distance from the seismometer to the earthquake.

b. Three seismic stations A, B, and C record the same disturbance. A and B are along an east-west line, with B located 450 km east of A.

Station C is 280 km due north of B. Figure 5.4 illustrates the geography. The arrival times of the compressional wave at A, B, and C are respectively 113, 144, and 112 s after the disturbance. How far is the event from each of the stations? In what direction will the event be as seen from B? In a sketch show how the location of the earthquake can be geometrically constructed.

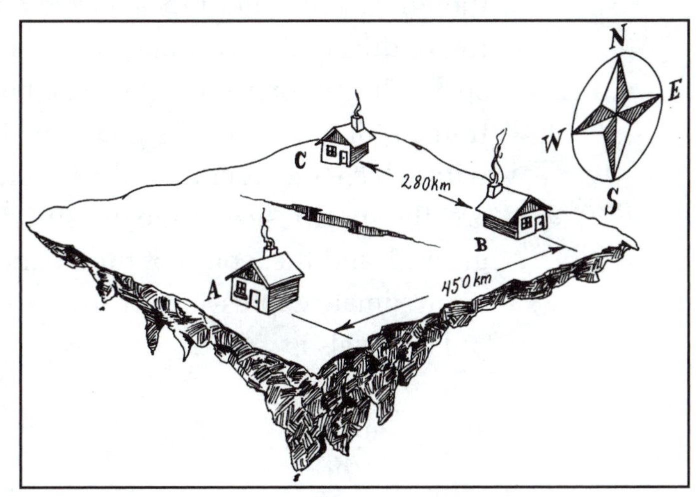

Figure 5.4

5.3.2 Disturbances which travel with some speed need not be destructive. They can even be a visual delight. Watch a squirrel running. A wave ripples along the animal's body and tail in such a way that the ripple appears to be stationary while the animal travels along it. The speed of the forward motion of the squirrel is quite different from the speed of the wave traveling along its body*. For the purposes of a physics problem, imagine that the squirrel runs through a stationary sinusoidal tube as in Figure 5.5. The squirrel has a length of 22 cm nose to tail and, as it runs, it occupies one complete cycle of the tube. The forward speed of the squirrel is 9.0 m/s. Any and all parts of the squirrel's body move up and down as if in simple harmonic motion.

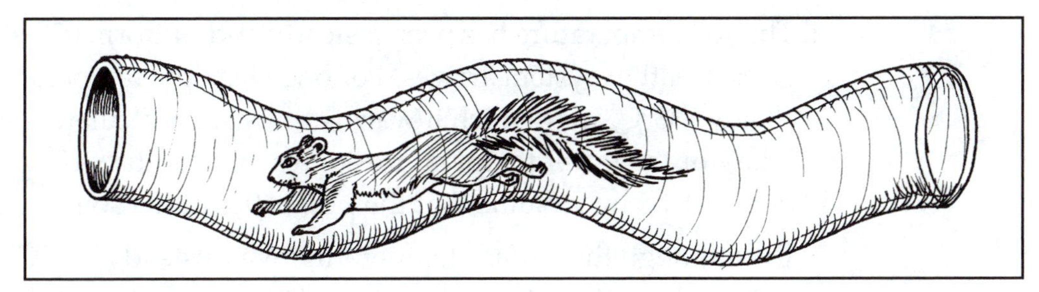

Figure 5.5

a. Calculate the frequency of the up-and-down motion.

b. What are the velocities of the disturbance and of the wave?

*For future reference keep this picture in mind as an analogy for the photon dealt with in later sections.

5.3.3 After a long period of extremely cold weather it often comes as a surprise that warmer temperatures do return. What seems strange though is that water pipes buried one or two meters under the ground freeze during the warming weather rather than at the peak of the cold spell. The reason for the delay is that the ground is extremely slow in transmitting the "cold" signal. The factors which determine the speed with which a temperature fluctuation is transmitted through the ground are: the specific heat c, the density ρ, the heat conductivity k of the ground, and the length of time T the unusually cold weather lasts. An approximate expression for the speed v at which such a temperature pulse travels is

$$v = \sqrt{\frac{2\pi k}{c\rho T}}$$

Another aspect of temperature fluctuations below ground is that the insulating qualities of the ground smooth out the severities of the temperature fluctuations. The dissipative (damping) effects are usually the dominant aspects of heat flow. The decrease in amplitude of the temperature change θ with distance x involves the same variables as the speed of heat flow.

$$\theta = \theta_0 e^{-\alpha x} \quad \text{where} \quad \alpha = \sqrt{\frac{\pi c \rho}{kT}}$$

The numerical values of soil properties are highly variable. The properties of ice are known and illustrate what can happen after a cold spell is over. For ice the values of c, ρ, and k are respectively 2.00×10^3 J/kg °C, 917 kg/m^3, and 2.20 J/s m °C.

a. The soil temperature has been a steady and uniform 0°C when an arctic weather system arrives. For one week the temperature is a steady –38°C, after which it warms up again. Calculate the time it takes for the lower temperatures to reach a depth of 10 cm and a depth of 1.5 m. Assume that the properties of the soil are those of ice.

b. Even though the surface temperature was a steady –38°C it is does not get that cold below the surface. The temperature difference between the soil below ground and the surface was θ_0. When the deepest cold finally reaches the 10 cm and 1.5 m mark, what will the temperature drop be?

c. Repeat the calculations of parts a and b based on a cold snap of –38°C which lasts only 24 hours.

5.4 SINUSOIDAL WAVES

An event at a source can result in a similar event occurring at some later time, and farther away. Mathematically this can be expressed as $y = f(x - vt)$, where y is the effect noted, x is the position along the direction the disturbance travels, v is the speed with which the disturbance travels in the medium, and t is the time delay. For example, the sound of someone chopping wood can be heard by a person standing close by or farther away. However, when a person watches the work from a distance, the chopping noises are heard while the axe is being lifted rather than when it makes contact with the block. The mathematical expression above ignores the fact that most media slowly absorb energy from the disturbance and that the amplitude drops significantly with distance.

Things become more interesting and mathematically most approachable if, instead of being a pulse-type event, the event is a repetitive one. It is particularly interesting if the repetitive event obeys the laws of simple harmonic motion and if the absorption of energy with time and with distance can be completely ignored. Scientific progress and improved understanding are achieved by getting a good understanding of the simple ideas and then adding in the complexities step by step to approach the realistic situations.

5.4.1 The common event which most closely obeys the idealized form of the mathematical wave equation is a lake with a gentle breeze blowing across it. Waves ripple and march along in straight rows.

The waves on a lake on a lovely summer day have a height of 11 cm peak to trough. At each given moment the distance between successive peaks is 25 cm, and the wave travels at a speed of 8.1 km/h north to south.

a. At the time and position arbitrarily designated as $t = 0$ and $x = 0$ the displacement of the water surface has its greatest negative value. Write an expression for the displacement of the water surface at $x = 0$ as a function of time t.

b. A snapshot is taken showing the positions of the peaks and troughs along the x direction. Write an expression for the displacement of the water surface as a function of x. (The x coordinate is lined up along the north-south direction.)

c. The troughs and peaks of the wave travel in the $+x$ direction. How long will it take for a trough, which was at $x = 0$ at time $t = 0$, to get to the position $x = 9.8$ cm? Use this information to modify the expression written in response to parts a and b to be valid at this new position and any later time t.

d. Focus your attention on the specific trough which was located at $x = 0$ at the time $t = 0$. Show that it obeys the condition $y = f(x - vt)$, i.e. that the displacement of the surface does not change when $x - vt$ remains the same. What in fact is the function f?

5.4.2 When an earthquake occurs at the ocean floor there is a tremendous amount of energy transformed into a momentary shifting of large amounts of water above it. The displacement of the water creates a pressure pulse which travels at the speed of sound in water. If and when the pulse reaches shore, the energy is concentrated in the shallow water of a bay, instead of being distributed over thousands of meters of water depth. This concentration can create extremely destructive waves called tsunamis.

The background information required to do some quantitative calculations on the tsunami phenomenon is as follows. First there is the fact that the speed v with which water waves travel in the oceans depends on the depth h of the water. The equation is

$$v = \sqrt{gh} .$$

The second relationship that must be accepted at this stage is that the total energy E carried in one complete wave depends on the amplitude of the wave A, the wavelength λ, the density of the liquid ρ, the acceleration of gravity g, and the width of the wave w. That equation is

$$E = \frac{1}{2} A^2 \lambda g \rho w .$$

The first equation predicts that a wave travels more slowly in shallow water than in deep water. One result is that the wavelength will decrease. The second prediction is that the same energy E is now concentrated in a shorter space λ and therefore the amplitude A must increase to compensate. Finally consider a funnel-shaped bay. The wave, as it enters, has a width w. The energy again concentrates as the bay narrows, with the consequence of a further rise in the amplitude A.

An earthquake has occurred below the ocean and creates a wave only 30 cm high but with a wavelength of 1.0 km in water 5.0 km deep. This wave approaches the shore along the west coast of Vancouver Island. The depth of the water decreases from 5.0 km to 20 m.

a. Calculate the velocity of the wave in the deep ocean and near the shore.

b. Calculate the wavelength and the amplitude as the wave reaches the shore.

c. The wave now reaches an inlet with high rock walls and a width of 1.0 km at the inlet. The inlet narrows to 30 m at the end. The depth remains uniform at 20 m. What will the amplitude of the wave be as it hits the end of the inlet?

The skills of skiing and surfing cannot be learned in a physics course, but the principles which lead to an understanding of skiing and how a surfer rides the waves can be obtained from a knowledge of the wave equations without excessive mathematics.

5.4.3 First some ideas based on skiing, in particular how the skier controls his speed. Given uniform snow conditions on a slope, the highest speeds are achieved on the steepest slope. The actual speed the skier achieves depends, among other things, on the quality of the skis, the snow conditions, and air resistance. The speed (actually the velocity) has three components: forward, downward, and possibly sideways. When the skis are pointing straight down along the slope (schussing), the skier will have higher downward and forward speeds than when the skis are pointed partly to one side (traverse position). In the traverse position the downward and the forward speeds are reduced and a sideways motion is introduced.

a. The ski hill at Mt. Trigonometry is shaped like a sine curve. The height is 322 m and the horizontal distance between top and bottom is 292 m along the expert run. A different part of the same ski area for intermediate skiers has the same sine curve shape but here the horizontal distance between top and bottom is three times as large. Finally, for the beginners, a separate area has been developed on Mt. Secant. For strange reasons this too is shaped like a sine curve but is only 75 m high with a horizontal distance between top and bottom of 405 m. Figure 5.6 is an illustration of the ski area. Calculate the slope in degrees of the steepest part of the ski hill for each of the three runs mentioned.

b. To a reasonable approximation the speed which a skier attains on a slope is proportional to the sine of the angle of the slope. Air resistance and other types of friction are assumed to be the controlling factors in this approximation. Ignore the fact that it actually takes some time for the skier's speed to react to changes in the slope. Sinuous Sylvia schusses down the expert run and attains a maximum speed of 92 km/h. Calculate the speeds Sylvia can reach on the intermediate and beginner hills using the same skiing technique.

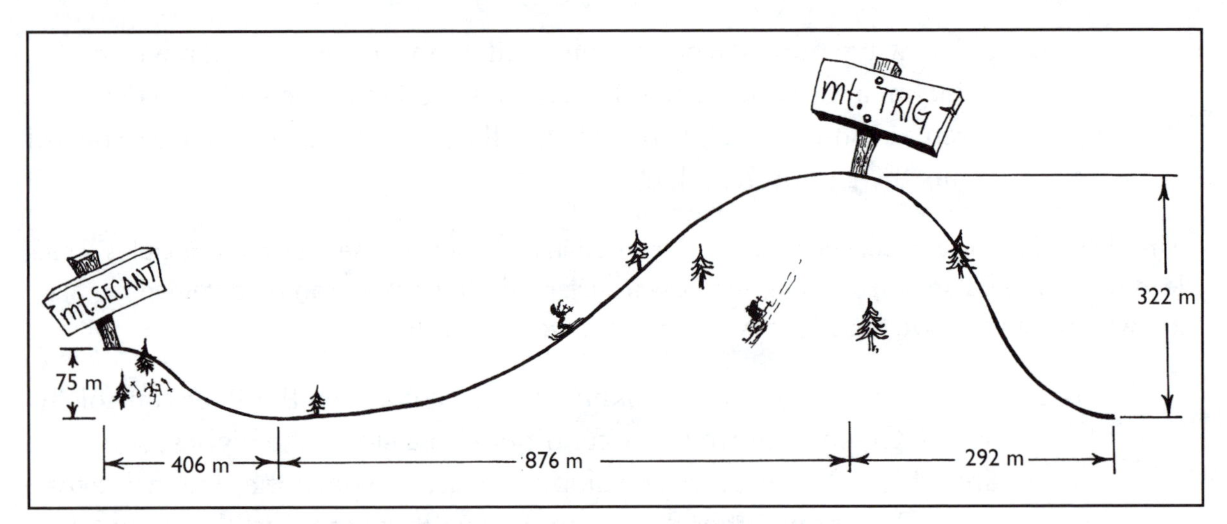

Figure 5.6

c. Because Sylvia is following the contour of the hill as she goes downhill, her speed has vertical and horizontal components. Calculate these components at the steepest sections of the three ski slopes.

d. Timorous Tim has managed to stay on his feet on the beginner's slope, even schussing straight down without regard to the safety of other skiers. He eventually learns to control the direction of his skis and discovers that by pointing his skis sideways instead of straight down he can actually control his speed. At what horizontal distance from the top will he have to start skiing toward the side to avoid speeds over 20 km/h? Assume that he would attain the same speeds as Sylvia under the same conditions.

e. Even Sinuous Sylvia checks her speed occasionally by going sideways. Derive an expression that relates the speed of descent to the angle of the slope α and to the angle θ Sylvia makes with the fall line (i.e. the direction of steepest descent). Specifically, at what angle with the fall line should she point her skis at the steepest part of the beginner's hill to slide along there at 20 km/h?

f. The forward component of the skier's motion needs further analysis before we can take a look at the sport of surfing in the next problem. In part e you were asked to calculate the speed of the skier on the hill as a function of the slope of the hill and the direction the skier takes with respect to the fall line. Now go the further step and derive expressions for the three individual components of the skier's motion, downward, forward (i.e. the horizontal component of the velocity in the direction of the fall line), and sideways (i.e. the horizontal component of the motion perpendicular to the fall line).

g. Sylvia achieves a certain forward speed component (which you can calculate) when she reaches the first section of the hill where the slope is 30° while she points her skis straight down the fall line. This forward component is important for her to maintain because she is trying to learn something about the art of surfing for the coming summer. At a steeper section of the hill she points her skis to an angle of 15° away from the fall line to achieve some sideways motion. What must be the slope of the hill at that point to retain her previous forward speed? What sideways speed will she have at this point?

5.4.4 During one particularly cold winter Sylvia flies to Hawaii. She discovers that surfing bears some resemblance to skiing. On calm days the waves of the Pacific Ocean roll in looking like low, gently moving sine waves. If only the waves were frozen in place instead of continually moving, Sylvia would have no trouble skiing down from crest to valley. Even on the moving waves the same basic story holds as on the ski hill: the steeper the slope, the higher the speed of the surfer. The surfer can also achieve forward, sideways, and downward components of a velocity. A little bit of thought shows that even though water waves are very small compared to ski hills, long rides are possible. First of all, even though the waves move forward, this can just be matched by the forward motion of the surfer sliding down the wave. Second, waves have a built-in ski lift. The water rises at the front edge of the wave and carries anything upward that floats on the water, including the surfer. In other words, if the surfer has just the right forward component of velocity, equal to that of the moving wave, and the lift of the water is just enough to compensate for the downward component of the surfer, then she can stay at a place of constant phase in the wave and ride toward shore.

Keep in mind one other variable. When skiing on packed slopes the person's skis follow the surface of the almost solid snow surface. When skiing in deep powder snow or riding a surfboard in the waves, the angle of the skis or surfboard need not be the same as the slope of the surface; therefore there is additional control available.
WARNING REPEATED: The skills of skiing and surfing cannot be learned in a physics course!

Based on the general principles of wave motion and the ideas of skiing developed in the previous problem, we can draw some conclusions for the requirements of surfing.

A few numbers to start with to help put the sport in context for those living far from ocean shores. Surfing is done on boards long and wide enough to float with a person standing on them. The board is approximately 50% longer than the rider. The wavelengths of the water must as a consequence be many times longer than the boards. Lakes rarely develop these long wavelength waves. Ocean waves show barely a ripple far from shore, but the long-wave low-amplitude ripples in deep water become higher waves in shallow water (see Problem 5.4.2). The waves start like sinusoidal waves off shore, but close to shore the retreating water undercuts the advancing waves and produces breakers and foam. The speed of ocean waves decreases as the depth of the water decreases. For the purposes of this problem use a ballpark figure of 10 m/s. The amplitude of the waves can be well over one meter. The depth of the water must be several times the amplitude of the waves.

After such a long introduction try for some theoretical answers to theoretical surf and theoretical surfing.

a. Sylvia has paddled her surfboard a few hundred meters off shore to wait for the right wave to come along for the ride back. At this location the speed of the waves is 15 m/s, the amplitude of the wave is 0.50 m, and the wavelength is 18 m. As Sylvia bobs up and down in one spot waiting for the next one, how much time elapses between successive waves?

b. The same wave has slowed down to 10 m/s closer to shore. What will its wavelength be there, and what time will time elapse between successive waves?

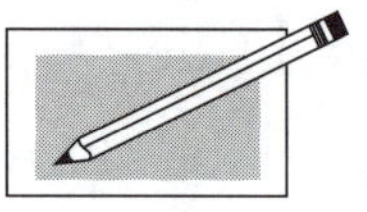

c. To ride the surf you need to move with the speed of the wave. To catch a particular wave you paddle as fast as you can and maneuver yourself onto the forward facing slope of the wave, point "downhill," and hope that the wave will carry you along from there. Explain why the wave must have more than some minimum amplitude to carry Sylvia (or anyone else) along. Would you expect there to be a wavelength dependence? At what section of the wave would you expect the highest forward speed? What will happen if the forward speed there exceeds the speed of the wave itself? It is possible to move faster than the wave without changing the location of the surfer with respect to a given phase of the wave? How can that be done?

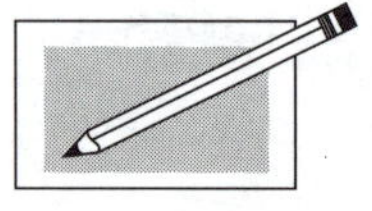

d. Expert surfers, unlike Sylvia, are not satisfied riding the smooth sinusoidal waves. The real excitement is to ride the incipient breakers, where the water just begins to curl up over the surfer's head (Figure 5.7). Explain why higher speeds of surfing are possible here than on a sinusoidal wave. Also try to explain how the slope at which the board is kept with respect to the water surface (e.g. the rear of the board can deliberately be partly submerged) adds further control over the surfer's motion.

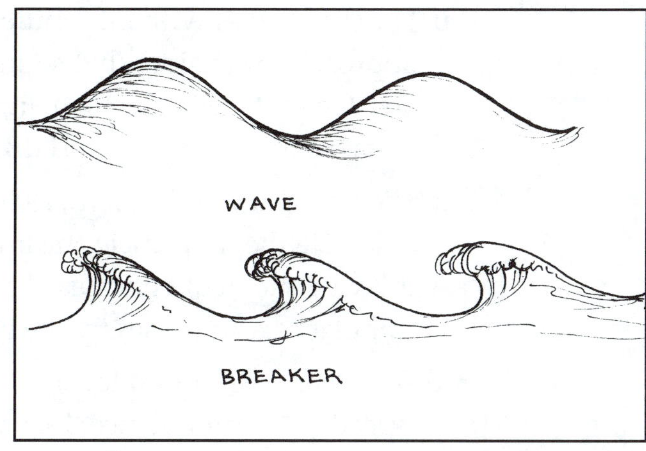

Figure 5.7

5.5 DOPPLER EFFECT

5.5.1 The Doppler effect can be demonstrated by whirling a tuning fork on a string in a horizontal circle. Suppose the length of the string is 1.24 m, from the center of the circle to the tuning fork. The length of the tuning fork is to be neglected, but its frequency, the note "A," is at 880 Hz. The tuning fork travels in its circle at a speed of 7.6 m/s moments after being struck to sound its note.

a. Determine the frequency(ies) an the observer will hear.

b. What frequency(ies) will the demonstrator hear as he whirls the tuning fork around over his head?

c. The frequency of the sound heard by the observer changes on a regular basis. Determine the frequency of this change.

5.5.2 The ear can readily distinguish between rapidly approaching and rapidly receding objects. For example, an approaching ambulance, siren screaming, becomes louder as it approaches and quieter after it has passed an observer. Less noticeable is the change in pitch explained by the Doppler effect. For the following analysis, assume that you are listening to the siren of an ambulance coming straight toward you, passing directly in front of you, and then going away from you. The ambulance travels at a speed of 73 km/h, has a siren which emits a steady tone at 512 Hz, and creates a sound level of 84 dB at a distance of 200 m.

a. Calculate the intensity of the sound at 200 m in W/m^2.

b. For street noise with intermittent traffic, it is a reasonable first approach to consider that sound intensity falls according to the $1/r^2$. law. On that basis, calculate intensity of the siren at 50, 4.0, and 1.0 m distance in both W/m^2 and dB.

c. The frequency of the siren, as heard by the observer, shifts suddenly. Calculate the frequencies heard by the observer. Sketch a graph showing the frequency heard as a function of the position of the ambulance as it passes by the observer.

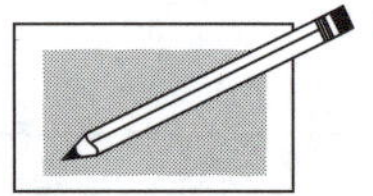

d. The change in sound level and the Doppler effect depend on the speed of the ambulance. Explain which of the two would be more practical to determine that speed.

NOTE: In practice neither is used. Police use radar or they measure the time it takes for a vehicle to go a measured distance.

The Doppler effect is used extensively in astronomy to measure the motion of stars. In acoustics the observed wavelength and frequency are calculated from the wavelength and frequency of the source by comparing the speed of the observer and the speed of the source with the speed of sound. The Doppler effect in astronomical observations is based on a comparison of the relative speed of approach or separation compared to the speed of light. With the speed of light being 299 792 458 m/s compared to the speed of sound at 330 m/s, it should be obvious that an object or an observer must move extremely fast to be able to detect wavelength changes in light or radio signals.

The precise mathematical expression for the optical (electromagnetic) Doppler effect looks quite different from its acoustical equivalent. However, as long as the speeds of the observer and/or the speed of the source are less than 5% of the speed of the wave, an approximation can be used which is valid for both optical and acoustic Doppler effect. The expressions become

$$f = f_o\left(1 - \frac{v_{rel}}{c}\right) \quad \text{and} \quad \lambda = \lambda_o\left(1 + \frac{v_{rel}}{c}\right).$$

In these expressions c is the velocity of the wave (sound or light) and v_{rel} is the relative speed between the source and the observer, to be considered positive when source and observer are separating and negative when they are approaching. λ_o and f_o are respectively the wavelength and the frequency of the source, while λ and f are respectively the wavelength and the frequency the observer measures.

5.5.3 Lawyers will go to great lengths to defend their clients. A lawyer tried to defend a client charged with running a red light. The defense the lawyer used was that his client had been fooled by the Doppler effect. He argued that it is well known that when approaching a light source there will be a shift in the perceived wavelength. His client had honestly mistaken the amber light for a green light. The judge accepted the argument but instead of fining the client for going through a red light, she charged him with exceeding the speed limit at $1.00 per km/h over the local legal limit of 50 km/h. With yellow (amber) assumed to have a wavelength of 577 nanometers (nm) and green of 560 nm, calculate the fine the judge charged.

5.5.4 Our knowledge of the stars and galaxies would be much less if we could not use the Doppler effect to measure stellar motion. In the laboratory one finds that a hydrogen discharge lamp has a strong emission at 486.1327 nm. Other emissions wavelengths are also present.

A careful study of the light coming from stars shows that hydrogen is present in stars but that the 486.1327 nm emission is shifted. Other characteristic emissions of hydrogen are shifted in the same way.

a. For a certain star the 486.1327 nm hydrogen emission is found to be shifted to 485.32 nm. Calculate the relative speed of that star with respect to the earth, assuming the earth to be stationary. Is the star approaching the Earth or moving away?

b. The Earth is in an orbit around the sun and travels at a speed of 29.7 km/s. Part of the year it is travelling toward the star, part of the time it is moving away, changing the Doppler effect. How much of the shift in wavelength in part a could be due to the motion of the Earth? Good astronomical instrumentation is quite capable of detecting this shift.

5.5.5 A large fraction (about 60%) of the stars that can be seen through a telescope are binary stars (two stars orbiting around each other in a way similar to the Earth orbiting the sun). In many cases the orbits are such that each star alternately approaches the Earth and recedes from the Earth. In some cases the orbit of the stars is such that one star regularly blocks the light of the other and we see a brief decrease in the the brightness of what otherwise appears to be a single star. An example of such a double star system is the "star" known as Algol in the constellation Perseus. Algol shows a brief decrease in brightness once every 2.87 days as the dimmer of the two stars blocks the brighter one. The two stars circle around their common center of mass with a total separation of 1.01×10^{10} m. The brighter one has five times the mass of the dimmer one.

a. Use the concept of center of mass to calculate the radii of the orbits of the two stars. Based on that result and on the time it takes for a complete an orbit, calculate the orbital speeds of the two stars.

b. Each of the stars will be alternately traveling toward the Earth and away from the Earth. A spectral line astronomers look for is due to the emission or absorption of hydrogen measured at 434.05 nm from a laboratory source. What are the longest and shortest wavelengths of this spectral line from Algol as seen on Earth?

NOTE: In astronomical work it is a combination of Doppler and brightness measurements that permits the calculation of the masses of the stars, not the reverse.

5.5.6 Galaxies are large collections of individual stars, but most are at such a great distance from Earth that telescopes pick up the light from galaxies as fuzzy spots. Individual stars can no longer be resolved. Still, some spectral lines can be recognized, particularly one due to calcium, which from an Earthbound source would be found at 397 nm.

a. A galaxy is moving away from Earth at 4200 km/s. At what wavelength would the calcium spectral line from the galaxy be detected at the observatory?

b. In 1929 Edwin Hubble correlated the distance of galaxies from the solar system with their Doppler shifts. It appeared that the farther a galaxy is away from us, the faster it recedes. The mathematical relationship is a very simple one,

$$v = d\,H$$

where v is the speed at which the galaxy moves away from us, d is the distance from us to the galaxy, and H is known as Hubble's constant. Although there are still arguments in the astronomical literature regarding the correct value of H, use H equal to 70 km/s per megaparsec in this problem. In SI units the astronomer's value of H translates into 1.0 km/s speed of recession for each 4.4×10^{14} km distance. How far away from Earth would the galaxy of part a have been at the time it emitted the light we are detecting now?

c. The light year is a non-SI unit astronomers find useful. It is the distance light travels in one year. Calculate the distance represented by one light year, and calculate the distance to the above galaxy as expressed in light years. Note that this is numerically equal to the time it took for the light from that galaxy to arrive at the telescope on Earth.

d. We now see the galaxy where it was millions of years ago. Assuming the galaxy continues to move away from Earth at the speed given in part a, how much farther away from us is it now, at the time when its light has reached us?

5.6 STANDING WAVES

Musical instruments are created for the purpose of giving pleasure to the listener. A single, isolated violin string of length L can be made to vibrate at its fundamental mode with a wavelength λ equal to twice the length of the string. In addition any number of overtones can be created whose wavelength λ can be specified by the equation

$$n \lambda = 2L$$

where n is an integer. The frequency of the vibration in the string can then be calculated from the speed of propagation of a wave in the string, which in turn depends on the tension in the string and its mass per unit length. Anyone playing a string instrument is familiar with the tuning (adjusting the tension) required to achieve the correct note values from the instrument. A string vibrating by itself is hardly audible. In order to make music the string is mounted near a sounding board which is a characteristically shaped box which we identify with a violin, a guitar, or a sitar. The role of the vibrating string on the box corresponds to the role of the playground supervisor pushing the child on the swing (see Problem 2.2.2), while the air in the sounding board corresponds to the child on the swing. However, there is one fundamental difference between the child on the swing and the sounding board. The child swings back and forth at one, and only one, frequency. In contrast, the sounding board has been designed to allow the material and the air in the box to vibrate simultaneously over a wide range of frequencies. In addition, once the board has absorbed energy from the string, it is far more efficient at setting the air around it in motion in order to send the tones to the listener. The shape and the material of the box are critical in an acoustic instrument to send out the desired combination of fundamental frequency and overtones to please the ear. The principle is the same in other instruments. For example, in wind instruments the vibrating lips and/or a vibrating reed replace the string and characteristically shaped tubing replaces the sounding board.

5.6.1 A guitar has 6 strings. All are the same length as stretched between the bridge at the bottom and the screws at the top of the instrument. Each string is differently constructed and each has its own tension adjustment. Therefore, by appropriate adjustment of the tension, you can create any fundamental frequency required to cover the range of human hearing. The mix of higher frequency overtones permitted by the same length and tension of the strings and properties of the sounding board characterizes the quality of the instrument. The length between the bridge and the screws over which the string can vibrate freely has been standardized at 0.628 m.

a. What range of tensions would be required to cover, as a fundamental, the frequency range from 164.8 Hz to 659.2 Hz for a string of 5.28 g/m. This weight of string is most often used for bass notes.

b. Six strings are used on most guitars to cover the frequency range mentioned in *part a*. Calculate the force, in newtons, the guitar neck and body would have to withstand if all six strings had a mass of 5.28 g/m length. The fundamental frequencies of the strings must be 164.8 Hz, 220 Hz, 293.7 Hz, 392 Hz, 493.9 Hz, and 659.2 Hz. Also express that force in terms of equivalent kilograms of mass hanging from the guitar body.

c. To reduce the strength requirements on a guitar each of the strings are made of different diameter material and often of different materials altogether. Steel, nylon, and catgut are or were the materials most commonly used. The material is chosen partly to reduce the stress on the guitar frame, partly to ensure tonal quality, and partly to ensure that the skin on the fingers can tolerate it without being cut. Suppose then that the heaviest string at 5.28 g/m is used for the lowest note, E at 164.8 Hz, and that the same tension is to be used on all strings. Calculate the mass per meter length of the other strings which are tuned to A, D, G, B, and a higher octave E. The fundamental frequencies of these notes must be respectively 220 Hz, 293.7 Hz, 392 Hz, 493.9 Hz, and 659.2 Hz.

d. With these realistic strings tuned and in place, express the total force of the applied tensions in terms of equivalent kilograms of mass hanging from the guitar body.

e. The pressure of the finger against the string and against the fingerboard of the instrument maintains the tension at its previous value while the length of the string that vibrates as a unit is shortened to where it touches a fret. The D string has been adjusted to give a fundamental frequency of 293.7 Hz. How far from the end of that string (from the screw) should the frets be located to obtain the notes designated as F, G, A, and B, corresponding to frequencies of 349.2 Hz, 392, 440 Hz, and 493.9 Hz when the finger presses the string to the appropriate fret?

5.6.2 A lecture bench demonstration consists of a tuning fork and a short length of wooden tubing made into a resonator. Both have been constructed to have the same fundamental frequency. The demonstration consists of tapping the tuning fork and hearing the tone it is designed for, and then touching the tuning fork to the wooden resonator. The result is that a tone is heard which is much louder but still at the frequency of the tuning fork.

a. The resonator acts like an organ pipe which is closed at one end. Calculate the length of the resonator required to match the frequency of a 440 Hz tuning fork.

b. While the demonstrator is looking away, a student squirts a dose of helium gas into the resonator. Explain what you expect would happen to the operation of the tuning fork and the resonator.

c. Predict the frequency of a tuning fork required to bring it into resonance with the helium- filled resonator.

5.6.3 There are optical analogs to the standing waves in an organ pipe. There are standing waves in the Fabry-Perot interferometer and in laser diodes. The Fabry-Perot interferometer consists of two highly reflecting mirrors, usually just separated by air. Its standing wave characteristics are used to select or measure wavelengths to high precision. The laser diode is a crystal that emits light when electrical energy is supplied. The light that comes out is of high spectral purity and is highly directional. Inside the crystal (the cavity) there is a standing wave of electro-magnetic energy (the laser light) because two reflecting mirrors on opposite sides of the crystal act like the ends of the organ pipe. The separation between the mirrors is one millimeter or less. The wavelength emitted by one type of laser based on gallium arsenide is 810 nm. The index of refraction in the cavity region of 3.54.

a. For reasons connected to the gallium-aluminum arsenide material, the output wavelength of these lasers is restricted to within a few nanometers of 810 nm. The length of the cavity, 1.00 mm, is far from the correct size for the fundamental frequency of a standing wave in the cavity. Use the given index of refraction of the material in the cavity to calculate the frequency and wavelength of the laser light while it is still in the cavity.

b. Assume that the cavity and the light wave inside the cavity behave as sound behaves inside an organ pipe closed on both ends. Based on wavelength in the cavity and length of the cavity, calculate approximately which overtone the light wave in the cavity represents.

c. By how many nanometers would the wavelength of the light have to shift if the overtone number calculated in *part b* were increased or decreased by one? Close scrutiny of the light coming from laser diodes shows that many closely spaced wavelengths come from the laser as shown in Figure 5.8. These are individual overtones as you have just calculated.

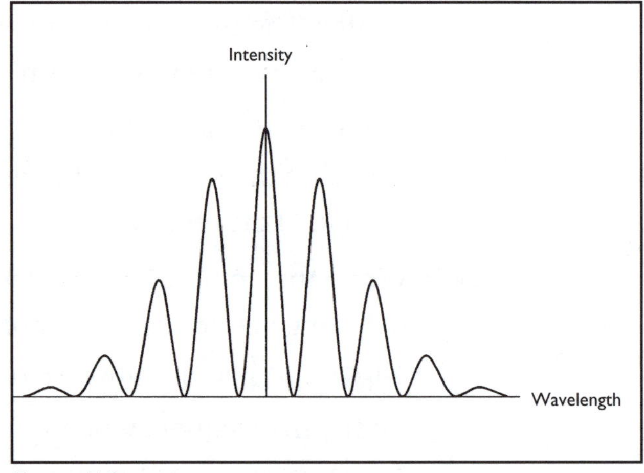

Figure 5.8

5.7 BEATS

5.7.1 Musical instruments must be tuned in order to emit the correct note values. For example, if two instruments play the same note and due to poor adjustment one instrument is slightly "out of tune" compared to the other, the result is unpleasant sounds. On string instruments, such as violins, guitars, and sitars, the same note can be obtained from different strings or the note from one string may have overtones in common with a different note struck simultaneously on a different string. If these different strings have not been properly tuned, then beats can appear and unpleasant sounds can result.

The D string (fundamental frequency 293.7 Hz) of a guitar is to be tuned to the A string (fundamental frequency 220 Hz). The A string is assumed to be correctly tuned, having been tuned to a piano or because it sounded correct to the ear.

a. One of the strings is allowed to vibrate in its fundamental mode, the other is shortened, by pressing the string against the correct fret, to sound the same note. Which string must be shortened, and to what length, such that the two strings create the same fundamental note when the tension in the D string has been properly adjusted?

b. A person with good hearing can detect beats between the notes from two strings if they are not precisely in tune. At an early point in the tuning stage a beat frequency of 5 Hz can be detected because the tension in the D string is not quite correct. In that situation, what is (are) the frequency (ies) of the D string?

c. The string is slowly tightened and the beat frequency has shifted to 1 Hz. Calculate the change in tension in the string.

It is often simpler to compare two objects than to measure each one individually to high precision. A familiar situation where a comparison is simpler than a measurement is to stand two people back-to-back and to measure their difference in height instead of measuring each person's height individually. When two frequencies are to be compared the primary tool is the detection of beat frequencies.

5.7.2 A client brings a quartz crystal watch to a jeweler because the watch gains an unacceptable 45 seconds per day. The jeweler adjusts the tuning of the crystal oscillator and then must check to see that the adjustment is correct.

In days of yore the jeweler would have carefully set the time on the watch to his standard clock and then waited for a day to observe the difference between the readings on the two time pieces. A further adjustment and another 24 hours may have been needed to achieve an adjustment to an accuracy within one second per day. The modern jeweler uses an electronic device that can detect the vibration of the crystal in the client's watch. The device then compares the frequency of the watch to a standard oscillator which runs at a frequency of 32 000 Hz. Fortunately all quartz watches use crystals within a small tolerance of the same frequency and can be adjusted to match that frequency. Electronics mixes the standard frequency with the frequency of the watch under test and the frequency is measured.

a. If the watch is ahead by 45.8 seconds per day, calculate by how much the frequency of the quartz oscillator of the watch is in error. What beat frequency will be detected against the standard? Explain what difference there would be if the watch lost 45.8 seconds per day.

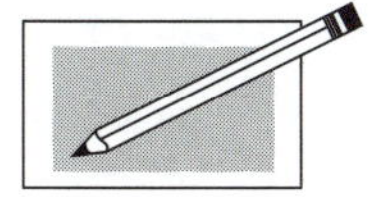

b. The jeweler has made an adjustment and now finds a beat frequency which fluctuates between zero and two herz. Explain the causes for such a fluctuation. Is it worthwhile to attempt a better adjustment? How long does it take to obtain enough data to determine these beat frequencies?

NOTES

6 ELECTRICITY AND MAGNETISM

6.1 ELECTROSTATICS

6.1.1 It is an unpleasant task to cut polystyrene bead-board. Small crumbs of the material in the form of beads stick to everything, including the knife or saw. Experts avoid the formation of the crumbs by using a hot wire instead of a sharp tool. Still, the nuisance of the foam crumbs deserves a few calculations. The stickiness of the foam beads is due to electrostatic forces. The friction while cutting the foam-board creates a charge-unbalance just like the build-up of charge that occurs when walking on a rug or while combing your hair in a room with very dry air.

To set the stage imagine your hands and arms covered with small polystyrene beads after insulating a wall. With one hand you pluck some beads off your clothing, you then try to shake them off your hands only to have the beads immediately fly back onto your clothing and reattach themselves. The question shows how to determine the magnitude of the forces, charges, and potential differences involved.

The specific gravity (density) of polystyrene bead-board is 24×10^{-3} g/cm^3. The beads released while cutting the sheets are typically 4 mm or less in diameter.

a. Begin by applying Coulomb's Law. One bead, 1.22 mm in diameter, is held stationary. An identical second bead passes by below. When the separation between the beads decreases to 0.95 cm, the second bead defies gravity and flies upward to attach itself to the first bead. First calculate the electrostatic force of attraction required on the second bead to counteract the gravitational force. Second, calculate the charge on the beads assuming that they have an equal but opposite charge.

b. For the situation described in *part a*, calculate the electric field and the electric potential at the surface of each particle due to its own charge.

c. The situation described, of just two small particles attracting each other, is unrealistic. More likely is a situation with a large, relatively flat surface, like your shirt, attracting a large number of small particles. Suppose the particles have the same size and charge as calculated in *part a,* and that the flat surface has the same electric field as calculated in *part b*. Explain why the particles will reattach themselves to your the shirt over a distance of many centimeters.

6.1.2 There have been attempts to explain the gravitational attraction between objects by a minute charge imbalance. In such a scheme, Newton's Universal Law of Gravitation becomes just a special case of Coulomb's Law. The primary aim of the hypothesis was to explain the origin of the force of attraction between Sun and Earth required to keep the Earth in its planetary orbit.

Suppose the Sun is positively charged. Suppose also that the net electrostatic charge on the Sun and on the Earth is in proportion to their masses and that there is no attraction between these objects other than the electrostatic one.

Solar System Data

	Sun	**Moon**	**Earth**
Mass	1.9891×10^{30} kg	7.348×10^{22} kg	5.9742×10^{24} kg
Radius	6.96×10^{8} m	1.738×10^{6} m	6.378×10^{6} m
Orbital radius		3.8440×10^{8} m	1.496×10^{11} m
Orbital period		27.3 days	365.25 days

a. Based on the above suppositions, would the Earth carry a net positive or negative charge? Calculate the net charges required on the Earth and on the Sun to keep the Earth in its current orbit.

b. If the hypothesis works on the Earth-Sun system, then the same hypothesis must be valid for the Moon-Earth system. Calculate the electric charge required on the Moon and on the Earth. Will the charge on the Moon be positive or negative?

c. Again based on the electric charge hypothesis, calculate the electric field and electric potential at the surface of the Earth strictly due to its own charge.

d. Still based on the electric charge hypothesis, calculate the electric field and electric potential at the surface of the Sun strictly due to its own charge.

e. In the days of the Apollo program, the manned lunar exploration spacecraft were first placed in Earth orbit, they were then boosted into a trajectory which intersected with the path of the moon, and finally they were slowed into an orbit around the moon. After the lunar orbit was achieved, the lunar lander was used. What, on the basis of the electric charge hypothesis, must happen to the charge on the Apollo spacecraft as it journeys to the moon and back?

6.2 CIRCUITS

6.2.1 Strings of Christmas lights are bought ready-made. An occasional adventurous spirit can still connect lamps in an unorthodox fashion. In this particular case the adventurous spirit is sufficiently cautious to use a 12 V storage battery and a large number of identical low wattage bulbs designed for 12 V operation. The circuit is shown in the diagram below. The lamps are labeled A, B, C, etc. List the lamps in order of brightness when the circuit operates. Justify your answer.

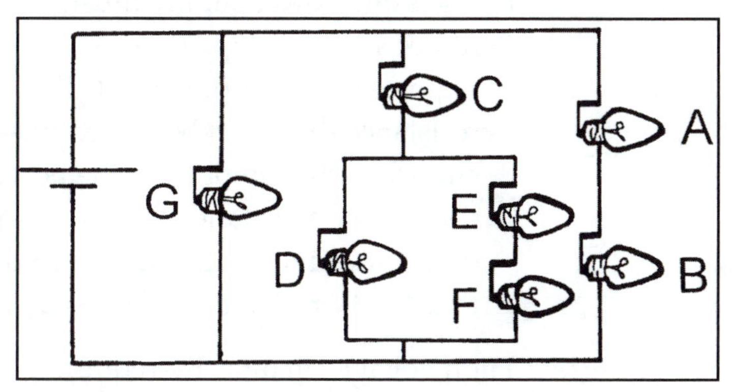

Figure 6.1

The mass and size of objects correlate reasonably well. Large objects are usually heavier than small objects, although it is easy to fool the eye by making fake barbells out of polystyrene foam and painting them to look like solid metal. The electrical resistance of an object however is completely unrelated to size. The resistance value of the small resistors that are found in radios and other electronic devices can only be determined by actual measurement or by a color code associated with four colored bands on the resistors. They all look like small cylinders with a wire coming out of each end. The physical size and shape of a 1/2 ohm resistor is indistinguishable from a 10×10^6 ohm resistor. The size of the resistor does mean something. The larger the size of the resistor, the more power can be dissipated in the resistor. In the cheapest and most common resistors a permitted power dissipation of 1/4 W is typical for the smallest physical size, 7 mm long, 2 mm in diameter. For a 2 W resistor the cylinder is 15 mm long and the diameter is 7 mm.

There is one more characteristic that the electronics technician learns to deal with. Although, at a price, various values of resistance are available, the most common ones have resistances that can be characterized as $1.0 \times 10^m\ \Omega$, $2.2 \times 10^m\ \Omega$, and $4.7 \times 10^m\ \Omega$, where m is an integer between 0 and 6. Other values required by circuits are patched together by parallel and series connections.

6.2.2 A typical power supply in electronic circuitry produces a highly stabilized 12 V dc to run the chips, transistors, lights, and loudspeakers.

a. A small red indicator light (LED: Light Emitting Diode) needs 20 mA to operate at approximately 0.7 V. Determine the resistance that must be placed in series with the diode to restrict the current to the diode from the power supply to 20 mA. What combination of standard resistors from the list in the last paragraph of the introduction should be used? Note that + or – 10% precision is adequate in such circuits. Is the power dissipation such that 1/4 W resistors can be used?

b. The same power supply must, on occasion, drive a small electric motor such as a tape transport of a recorder. Suppose this motor requires 6.0 V to operate at 0.45 A. As in *part a*, determine the resistance that must be placed in series with the motor to restrict the current to the motor from the power supply to 0.45 A. What combination of standard resistors from the list in the last paragraph of the introduction should be used? Is the power dissipation such that 1/4 W resistors can be used? How can that be dealt with?

6.2.3 High voltage circuits also require the use of multiple resistors. The resistors are also more likely to be ordered for the specific application rather than the choice of a handful of small resistors out of a set of drawers. The resistors are large in all dimensions—length and diameter—to have a large surface area for heat dissipation. They are also often immersed in oil or PCBs to prevent sparking along the length of the resistor.

An X-ray unit is designed to operate at 60 000 V, with one side kept at ground potential and the other side at lethal levels. The dc voltage level **V** must be monitored continuously (mostly by automated equipment). A standard way of monitoring **V** is to

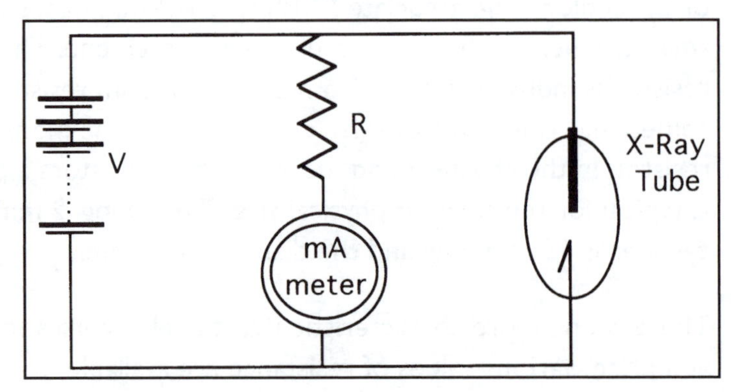

Figure 6.2

have large resistor(s) **R** in series with a milliampere meter **mA** functioning as a voltmeter across the line, as shown in the diagram.

a. The meter is factory adjusted to have a scale from 0 to 1.0 mA. For simplicity only the scale of the meter is changed. 0 remains 0; 1.0 mA now indicates 100 kV. Values in between are in the same proportion as before. Determine the required value of R such that the meter correctly indicates the applied voltage V.

b. As stated in the introductory paragraph, it is dangerous to have a high voltage applied over a short distance. Immersing the resistor in oil helps to decrease the danger. Suppose that 20 kV is the maximum recommended voltage drop across any resistor in this circuit and that on occasion a brief voltage spike of up to 100 kV can occur. What arrangement of resistors should be substituted for the single R indicated in the diagram?

c. Calculate the power dissipated, at the operating voltage of 60 kV, in each resistor and in the total string of resistors.

6.3 OHM'S LAW

Electrical wiring in North America has its share of jargon and non-metric standards. For example, one speaks of 110 V outlets as standard in the home. Where high power is required, 220 V outlets are used for large appliances, such as electric dryers and electric stoves. In reality all 110 V outlets provide power at 120 V. Some 220 V outlets provide power at 208 V, others provide 240 V. The outlets are designed to accept only specially shaped connectors to prevent accidents. In North America the thickness of wires is specified by a number in the American Wire Gauge (B & S) system. The number decreases as the thickness of the wire increases. Flexible electrical cords on lights and radios are usually made with #18 gauge wire and are acceptable for currents up to 5 A. Hidden wiring in the walls of a house is done with #14 gauge wire or lower gauge wire. The cable bringing power into the house is even thicker. A few selected handbook values of the electrical resistance for single #18, #16, #14, and #12 copper wires are 19.3 Ω/km, 12.14 Ω/km, 7.63 Ω/km, and 4.801 Ω/km with cross-sectional areas of 0.8231 mm^2, 1.309 mm^2, 2.081 mm^2, and 3.309 mm^2 respectively.

6.3.1 In northern climates automobiles sprout external electrical connectors in the winter and cars parked on the street appear to be connected by umbilical cords to the nearest house. Although some car owners go so far as to install electrically heated seats, most are satisfied with an electrical heater in the engine block to keep the oil flowing freely.

A standard version of such a block heater consumes power at a rate of 800 W at the 120 V of the house electrical outlets.

a. Calculate the resistance of the block heater

b. Calculate the electrical current in the heater when it is in use.

c. Many cars are parked on the street and require extension cords to reach from the house electrical outlet to the block heater. A distance of 25 m is not uncommon. At the hardware store the choice is between a cable made from #16 gauge wire, which should be just about adequate (particularly in cold weather) for the required wattage, and a cable made from #14 gauge wire, which is actually recommended for the purpose but costs considerably more. Calculate the resistance of a 25 m length of #16 cable and then calculate the current through the cable when it is used to connect the block heater to the 120 V outlet.

d. Will the engine block still get 800 W at the end of the #16 wire cord? Calculate the heat energy dissipated in the block heater and in the cord leading to the car. If the voltage at the outlet in the house is 120 V, then what will be the voltage at the input of the block heater?

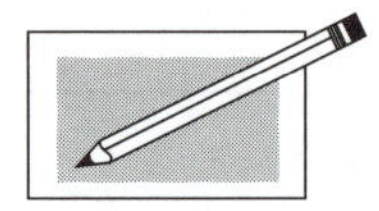

e. Repeat the calculations for *parts c* and *d* if a #14 wire cable is used. Comment on the effectiveness of the block heater when used with extension cords having higher or lower gauge wire.

6.3.2 A typical electric hedge clipper has a 1/4 hp motor (1 hp = 750 W). Calculate the power the clipper gets from a direct connection to a 120 V outlet as compared with getting its power through 30 m long extension cords made from #18 gauge and from #14 gauge copper wire respectively. In each case determine the power loss in the extension cord.

In North America, the electrical outlets deliver power for most appliances at 120 V ac. In Europe and Asia, 220 V ac is the standard. The frequencies are also different. In North America 60 Hz is standard, while 50 Hz is the norm in Europe and Asia. Electrical equipment built for the North American market—razors, radios, computers—will be damaged when plugged into a European outlet. Because power is the product of current and voltage, the higher the voltage the less current needs to be drawn and, as a consequence, less power is lost in distribution as heat in the wiring. The higher voltage does have the drawback of increased danger from accidental electrical shocks.

6.3.3 Two devices are routinely carried by travelers. The electric shaver, requiring 10.4 W to operate, keeps certain excessive hair growth under control. A small electric iron, requiring 125 W, keeps clothing neat. There is no problem with these devices as long as the travelers stay on their own continent. Some manufacturers specializing in the travel market install switches on electric devices to allow their use on either voltage as long as the traveler has the correct adapter in order to fit his plug to the foreign socket. This problem asks you to explore some of the intricacies of intercontinental travel.

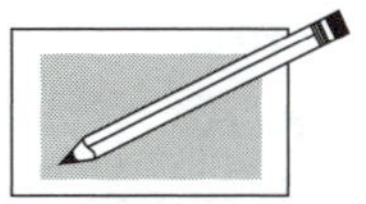

a. Begin with the European family traveling to North America. Discuss what will happen if they try to use their electric shaver and electric iron in 120 V outlets instead of the 240 V outlets for which they have been designed.

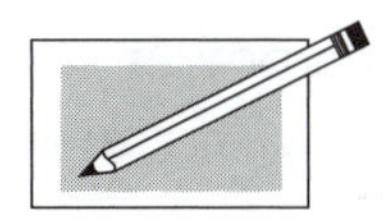

b. A North American family travels to Europe for a vacation. Discuss what will happen if they try to use their electric shaver and electric iron in 240 V outlets instead of the 120 V outlets for which they have been designed.

c. The North American family in Europe has three options. They can buy a new shaver and a new iron, buy a heavy ac transformer, or install a resistor in series with the device in question. The first option is financial rather than scientific and therefore will not be considered here. The second option is a valid scientific option but belongs in a later section. This leaves the third option. What resistance must be installed in series with each device to allow it to operate normally? What power will be dissipated in each resistor?

d. Discuss the options the European family has in North America.

6.4 OHM'S LAW AND MATERIAL PROPERTIES

Ohm's Law is based on experimental evidence for many materials in a great many situations. It is not a universal law because some materials do not obey Ohm's Law. Examples of these are in section 6.5. There are two aspects of Ohm's Law. The first aspect is that for any piece of material that obeys Ohm's Law a property called resistance can be specified. This means that it has been verified that the current through that piece of material is proportional to the applied potential difference (voltage). The second aspect is that the resistance of the piece of the material depends on its composition and a shape factor. For a wire, the shape factor is length divided by cross-sectional area. The composition of the material determines whether the wire can be used as a conductor, an insulator, or something in-between. This electrical property of the material is called the resistivity ρ. It must be experimentally determined. There are structural properties that help explain resistivity, but these are beyond the confines of this book.

The resistivity for superconductors is zero. The resistivity at 20°C for copper is $1.7 \times 10^{-8}\ \Omega$ m, $\rho = 22 \times 10^{-8}\ \Omega$ m for lead, within an order of magnitude of 100 Ω m for silicon , and approximately $10^{15}\ \Omega$ m for diamond. For all materials there is a temperature dependence of the resistivity and even for ostensibly the same material there can be significant variations in ρ due to small amounts of foreign matter (impurities) and due to changes in crystal structure. For example, two of the solid forms of carbon, graphite and diamond, have grossly different structure and resistivities.

6.4.1 Three different solid pieces are made in identical shapes. They are 5.2 cm long, have a square cross-section, and are 0.16 mm thick. A potential difference of 0.37 V is applied along the length of each piece.

a. One piece is made from copper, resistivity $\rho = 1.7 \times 10^{-8}\ \Omega$ m. Calculate the current flowing through the piece. How much power is dissipated in the volume of the piece?

b. The second piece has been made from a type of stainless steel with $\rho = 70 \times 10^{-8}\ \Omega$ m. Calculate the current flowing through the steel. How much power is dissipated in its volume? What voltage must be applied to achieve the same current as calculated in part a? What voltage must be applied to achieve the same power dissipation as calculated in *part a*?

c. The third object is made from rubber, $\rho = 2 \times 10^{13}\ \Omega$ m. Calculate the current flowing through this object. How much power is dissipated in its volume? What voltage must be applied to achieve the same current as calculated in *part a*? What voltage must be applied to achieve the same power dissipation as calculated in *part a*? Do these answers make sense?

NOTE: Good insulators often allow more current to pass than calculations predict. This is usually due to moisture or dirty finger prints on the surface creating a lower resistance coating. The currents flow along the contaminated surface instead of through the material itself.

6.4.2 Automobile batteries are large and heavy. Lead is a major component of automobile batteries which is one reason for the heavy weight. The reason for the large physical dimensions of a car battery is that it must store a lot of energy. All battery types have specific energy density characteristics that determine the volume of battery needed for the energy it has to supply. In addition to having enough total energy available, the automobile battery must also be able to deliver the energy in a short time (high power) which in turn implies a momentary high current. This too sets requirements for the size of the battery as the following calculations will show.

A typical, fully charged automobile battery has an output voltage of 12.6 V and should be capable of delivering 200 A to the starter motor for a few seconds. The current must be able to flow through the battery acid which has a resistivity ρ of approximately 3 Ω m. The laws of chemistry restrict the voltage output for a single battery cell to about 2 V, therefore all 12 V (nominal) automobile batteries consist of six 2 V cells connected in series.

a. Calculate the total resistance of battery and starter motor circuit if 200 A is to be permitted to flow momentarily from the fully charged battery.

b. Consider that half the resistance calculated in *part a* is in the motor and half the resistance is distributed through the battery. On that basis calculate the resistance in an individual cell of the battery.

c. The battery acid separating the plates in the battery fills a space of 0.3 mm. A special type of wet blotting paper in the acid keeps the plates from touching. Ignore the presence of this paper. If each cell of the battery consisted of just two large plates separated by the acid soaked blotting paper, calculate the area of the plates required to allow the current to pass. If the battery plates were square pieces of material, calculate the length of the sides of the square.

d. A quick look at a battery shows that the calculated dimensions from *part c* are inconsistent with a real battery. A sketch of three of the six cells are shown in Figure 6.3. The approximate outside dimensions of a real automobile battery is 20 cm high, 25 cm long, and 16 cm wide. The individual cell has the same height and width, but 1/6 the length of the total. In each cell the plates are interleaved as in the diagram. If each leaf is 14 cm by 14 cm, what if the effective area A of the plates to be used in the expression $R = \rho l/A$?

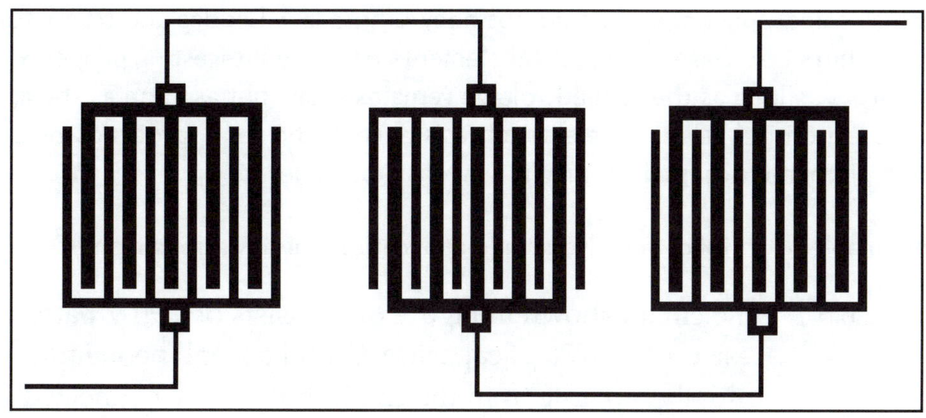

Figure 6.3

6.4.3 The heater wires in toasters and electric space heaters are made from high resistivity alloys. The alloys are also mechanically stronger than copper and they are stable against oxidation at elevated temperatures. An alloy based on nickel and chromium is widely used in heaters. It has a resistivity $\rho = 108 \times 10^{-8}\ \Omega$ m and a tensile strength almost four times greater than copper at room temperature.

An electric toaster has to spread the heat evenly over two slices of toast at a time. The heating element is wound from a long wire folded back and forth on each side of the bread slices to be toasted. The energy dissipation in the toaster is to be 980 W at 120 V. The single long heater wire has a length of 4.3 m.

a. Suppose the heater wire could be made from copper, $\rho = 1.7 \times 10^{-8}\ \Omega$ m. Calculate the required diameter of the wire. Ignore any changes in ρ with temperature.

b. More realistically, the wire is made from the nickel-chromium alloy. Again calculate the required diameter of the wire ignoring any changes in ρ with temperature.

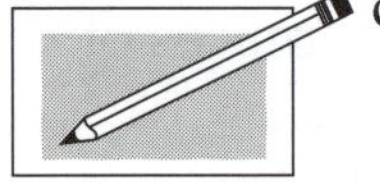

c. Based on the diameters calculated in *parts a* and *b*, give a rough estimate of the relative strengths of the copper and alloy wire. Justify your reasoning.

6.5 NON-OHMIC BEHAVIOR

Ohm's Law is not a fundamental law of physics. Rather it describes the behavior of a limited range of materials over a limited range of currents and temperatures. In that sense it is similar to Hooke's law as applied to the stretch of a spring—exceed the permissible force and the spring deforms permanently or it may break. Most metals and alloys, particularly those used in electrical circuits, closely obey Ohm's law. The contact points between dissimilar metals such as solder junctions often do not obey Ohm's law. Diodes and transistors also do not obey Ohm's law. Gases as electrical elements have the interesting property that they are good insulators as long as the applied voltage remains small, but as soon as the applied voltage exceeds a critical value, the electrical resistance of the gas approaches zero and an electrical spark discharge occurs. Lightning is a familiar example.

The problems to follow involve simple applications of deviations from Ohm's law.

6.5.1 The circuit shown in Figure 6.3 consists of a 90 V battery, a 12 MΩ resistor R, a 0.88 μf capacitor C, and a small neon lamp. The lamp is initially separate from the circuit but will be connected in parallel to the capacitor. This particular neon lamp is a perfect insulator as long as the applied voltage is less than 55 V. However, as soon as the 55 V limit is exceeded, the electrical resistance is less than 1 Ω. The current flowing through the lamp causes an orange flash.

Without the neon lamp, the operation of the circuit shown in Figure 6.4 is simply that of a capacitor slowly charged by the battery through the resistor. The potential difference across the terminals of the capacitor increases from zero to the voltage of the battery with the characteristic time constant $\tau = RC$. The initial behavior of the circuit with the neon lamp connected is the same. However, as soon as the potential difference across the capacitor reaches the 55 V limit, the lamp reaches its low resistance state and capacitor is shorted by the lamp, discharging all its accumulated charge. At the same time an additional current flows from the battery through the resistor through the lamp. The lamp regains its high resistance moments after the capacitor has been discharged completely. At that time the capacitor is again charged by the battery until the limit of the lamp is reached. The circuit, with the lamp in place, flashes at regular intervals with very low energy consumption. It is widely used as a warning signal.

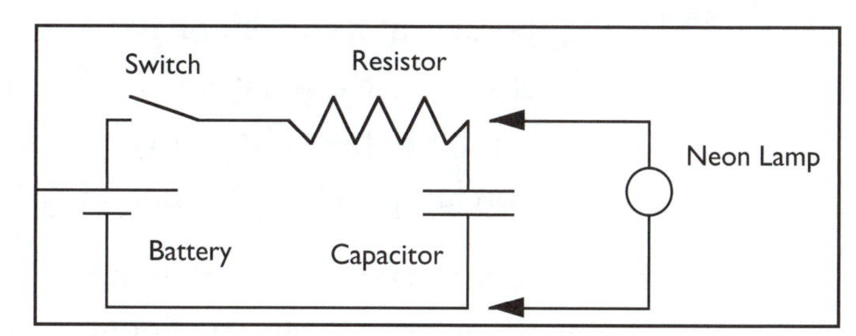

Figure 6.4

a. Consider the circuit without the lamp in place. At time t = 0, the charge on the capacitor is zero. Calculate the charge on the capacitor and the voltage across the capacitor after 0.500 s, 5.00 s, and 50.0 s.

b. Conversely, calculate the time it takes for the voltage across the capacitor to reach 70V and the time at which the voltage across the capacitor reaches to within 0.12 % of 90 V.

c. The neon lamp becomes highly conductive the moment the 55 V threshold is reached. What is the charge on the capacitor at that moment? Suppose the wiring inside the capacitor and the wiring leading into the gas of the neon lamp has a resistance of 0.47 Ω. Calculate the time required for 99 % of the charge of the capacitor to be discharged and calculate the average current from the capacitor during this time interval. While the capacitor discharges, the battery also supplies a current to the lamp limited by the resistor R. Calculate the current through the lamp from the battery and compare this with the current from the capacitor.

d. Calculate the time between successive flashes when the neon lamp is connected across the capacitor.

e. Calculate current, as averaged over many cycles, supplied by the battery. What will be the electrical energy drain on the battery?

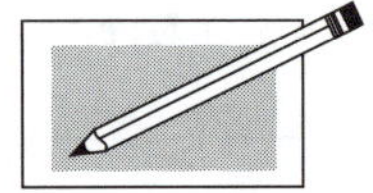

f. Speculate a little bit. A small light bulb as used in a flashlight operates at a current of 400 mA supplied by 3 V from two 1.5 V D cells in series. What power does this light bulb require to operate? Do you think this light source will be better or worse than the flashing neon lamp to alert the passerby? Explain your reasoning.

6.5.2 Some materials are innately non-linear in conductivity. A gas is an insulator for low electric fields, however, once the breakdown voltage is reached, the gas becomes ionized and conducting. Transistors and diodes are non-linear because inside these devices there are interfaces between regions of different impurities that act almost like one-way road signs for electrons. In this problem there is neither breakdown nor asymmetry; the non-linearity is due to heating effects.

The table lists a series of data points measured on a 60 W incandescent light bulb as the voltage across the lamp was slowly changed between 0 and 120 V. The applied voltage was measured in V_{rms}; the current was measured in I_{rms}.

a. Plot an I vs V curve based on the data provided.

b. Is I proportional to V as required by Ohm's Law? Check the low voltage data carefully.

c. Replot the data as log I vs log V.

d. There are two segments of the log-log plot that are close to being straight lines, but with different slopes. For these two segments, determine the slopes, $\Delta(\log I)/\Delta(\log V)$. Write the equations which best describe the I vs V relationship in the two straight line segments.

e. Why is it reasonable to expect the change of behavior? What would you have seen if you had done the experiment yourself?

Table of Data

Applied voltage (V)	Measured current (mA)	Applied voltage (V)	Measured current (mA)
0.144	7.8	24.8	218
0.204	10.6	28.1	230
0.288	15.5	32.7	246
0.433	22.9	39.9	270
0.621	32.0	48.8	298
0.862	42.7	55.4	318
0.945	46.2	62.7	339
1.17	54.8	68.6	354
1.65	70.5	74.4	370
1.66	71.2	80.8	386
1.99	80.2	85.5	397
2.50	91.5	90.1	409
4.52	121	94.7	420
6.44	137.2	102.1	436
9.74	156.4	108.4	451
12.22	167.9	110.4	455
13.6	173	119.8	475
17.54	190.3		

6.5.3 There exists an important class of electrical devices that better conduct electrical currents in one direction as compared to another direction. These are called rectifying diodes or rectifiers and they are vitally important in many electrical devices. They are the electrical counterparts of ratchets in mechanical systems and to arteries in blood flow.

Figure 6.5

The ratchet in a mechanical clock movement or in an automobile jack allows a gear to turn one way but also holds the gear in place when the force is released. The valves in arteries allow blood to flow forward, but close as soon as the blood pressure decreases while the heart prepares for the next push. Damage to the large valves in the aorta or in to heart itself are life threatening situations. It seems reasonable that the flow of the blood through the arteries has a pressure dependence: higher pressure likely means wider openings in the valves and consequently easier flow. In the same sense it should not come as a surprise that the resistance to current flow in an electronic diode varies with applied voltage.

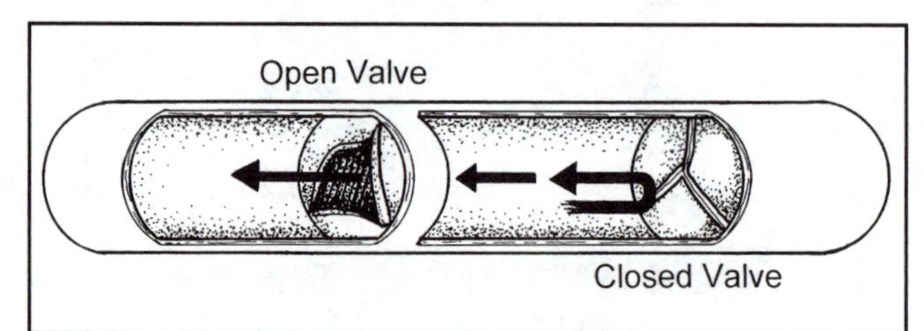

Figure 6.6

Theory predicts and practice closely confirms that the current-voltage relationship for the active part of a diode can be expressed as

$$I = I_o\left(\exp\left(\frac{AV}{T}\right) - 1\right),$$

where V is the applied voltage, I is the resulting current, $A = 1.16 \times 10^4$ K/C is an universal constant, T is the operating temperature of the diode in degrees Kelvin, and I_o depends on the design of the particular diode. The expression is valid over a wide range of voltages, positive and negative. In addition to the contribution of the rectification process, there is also an unavoidable resistance in series with the rectifying region that limits the current through the diode in the direction of easy flow. Diodes can be designed for high power applications where the series resistance is made small, or they can be miniaturized to be important elements in computer chips. They are used in applications requiring anywhere between hundreds of volts to less than a volt.

a. A given diode limits the negative current to $I = -1.2 \times 10^{-6}$ A at $V = -0.5$ V. Use this information to determine I_o and to calculate the current at –1.0 V and at + 0.5 V at 300°K.

b. When an electrically conducting material obeys Ohm's Law its resistance can be determined from $R = V/I$ or from R = (change in V)/(resulting change in I). For example, if two different values of current (I_1 and I_2)and voltage (V_1 and V_2) are given for a resistor, which obeys Ohm's Law, then the value of R can be obtained from V_1/I_1, from V_2/I_2, or from $(V_1 - V_2)/(I_1 - I_2)$. The calculated value of R will be the same in all three cases. Use the three values of the current for $V = -1.0$ V, – 0.5 V and + 0.5 V to calculate values for R from the expression $R = V/I$. Finally use the expression R = (change in V)/(resulting change in I) to obtain a value of R for the change in voltage from – 0.5 V to – 1.0 V.

6.6 MAGNETISM

6.6.1 A solenoid can be wound with many turns of thin wire carrying a small current or it can be wound with fewer turns of heavier wire carrying a larger current. The identical magnetic field can be achieved in either way.

A certain solenoid is wound around a cylindrical core with 2 500 turns of copper wire. The core has a length of 28.0 cm and a diameter of 2.54 cm.

a. Determine the required magnitude of the current in the coil to achieve a magnetic field of 3.15×10^{-2} T in the center of the coil.

b. The heat dissipated in the coil due to the current flow is to be kept at or below 12 W. Determine the minimum diameter of the copper wire and the total volume of copper used for the coil.

c. It was thought that fewer turns of thicker wire would permit a stronger magnetic field with less energy dissipation. Check out this idea by assuming that 900 windings of thicker wire are used.

Determine the required current in the coil and the volume of copper needed to achieve 3.15×10^{-2} T while still limiting the heat dissipated in the coil to 12 W.

d. The insulation on copper magnet wire is 0.02 mm thick. Assume the coils of *parts b* and *c* are tightly wound. Approximately how many layers of wire will there be in the two cases?

e. Calculate the inductance of the two solenoids.

6.6.2 Magnetic Resonance Imaging (MRI) machines for medical diagnostics require strong and extremely uniform magnetic fields over the volume of space in which the patient is located. The probing is done with echoes of applied microwave radiation from one small segment of the body at a time. Computers are used to construct the detailed image.

The required magnetic field strength is so large that only superconducting wiring in the magnet coil makes these instruments feasible. Superconducting wires at sufficiently low temperatures have not just low electrical resistance but actually zero electrical resistance. The coils are cooled to 4.2˚K with liquid helium as the refrigerant. In recent years materials have been discovered which are superconducting at temperatures as high as 130˚K—much easier to maintain than 4.2˚K. Unfortunately none of these so-called high temperature superconductors can tolerate the high current densities needed to generate the magnetic fields required for MRI.

A particular MRI machine for whole body diagnosis has a uniform magnetic field of 3.000 T over a cylindrical volume 2.00 m long and 1.20 m diameter. It is shaped like a large solenoid with windings which carry a current of 226 A. The windings, made from a NbTi alloy, are immersed in liquid helium in an highly insulating container to minimize heat leakage into the windings and frost to the patient.

a. Calculate the energy stored in the magnetic field of the solenoid. Although the magnetic field can never be completely confined to the interior of a solenoid, assume in your calculation that the field is negligible beyond the boundaries of the interior of the solenoid.

b. All electromagnets have an inductance. Based on the energy stored in the magnetic field and the current of 226 A, calculate the inductance L of the magnet.

c. Calculate the number of windings the magnet has, assuming it is just a standard solenoid. What length of wire does this represent?

d. Conventional copper wire recommended for 220 A has a cross-sectional area of 85 mm^2, a resistance of 0.203 Ω/km, and a mass of 756 kg/km. Suppose the MRI solenoid could have been made with coils of copper wire, calculate the resistance of the length of wire as obtained in *part c*. How much heat will be generated in the coil at 226 A?

e. If it were wound with copper wire, an MRI solenoid would need adequate cooling. How many liters of water at 12°C will be needed per minute if the water is allowed to heat to 23°C while passing through the coil?

The liquid helium in the MRI machine in our example must be replenished at a rate of 3 liters per day. Compare the cooling required by the real MRI machine to the cooling required by an MRI machine if copper wiring had been used. Liquid helium has a latent heat of vaporization of 2.72×10^3 J/ℓ.

f. The windings, refrigerant, and insulation of the superconducting MRI solenoid take up a space of about 10 cm thickness on the outside of the magnetized space. Estimate the thickness of the copper windings required if the solenoid had been made of copper. Ignore the additional space required for the cooling water and the electrical insulation between the windings.

The aspects of magnetism as experienced around the home are different from the aspects of magnetism as taught in the usual physics course. At home there are refrigerator magnets that hold notes to the refrigerator door: "phone George this evening" or "buy olives tomorrow." Magnets are useful devices that stick to iron walls, that can help find hidden nails or lost pins, and can, depending on orientation, attract or repel each other. Magnets are also used to find a direction, as with a compass.

In a physics course the emphasis is completely different. The properties of permanent magnets are too difficult to treat in a mathematically simple way, so current-carrying coils, called solenoids, are the favored sources for magnetic fields. Their fields can be most easily calculated. For the same reason, students are asked to calculate the force exerted by a magnetic field on some lone electron moving in a vacuum, rather than to find the push or pull of the solenoid on a piece of iron or on another solenoid.

In the following problems magnetism is dealt with in the way it was taught a century ago, when the use of magnets was more important than the logical development of the laws of electricity and magnetism. Central to this old-fashioned approach was the concept of magnetic poles, the idea that magnets obey a law like Coulomb's law—there are individual positive *N*[orth] and negative *S*[outh] magnetic charges (poles), that attract or repel according to a $1/r^2$ law. The complication in magnetism is that no experiment has ever found isolated *N* or *S* magnetic poles: By contrast, positive or negative electric charges are on electrons and ions in a vacuum. Positively or negatively charged metal spheres are also easily achieved on the lecture bench. Permanent magnets and solenoids always appear with equal amounts of positive and negative pole strength, separated by a distance determined by the size and shape of the device. In short, isolated magnetic poles are strictly fiction, but they can help to relate observations to a framework (model) of understanding magnetic forces.

The empirical laws that guide the model are
- $F = q_{m1}q_{m2}/4\pi\mu_o r^2$, as the magnitude of the force between two magnetic poles, and
- $B = q_m/4\pi r^2$ and $F = q_m B/\mu_o$ as the relations between magnetic pole strength and magnetic field.

To be consistent with SI units, the force *F* is measured in newtons, the distance from a magnetic pole *R* is measured in meters, and the magnetic field *B* is in webers/m^2 (= teslas,T). The magnetic pole strength qm then has units of webers (= T·m^2), and the constant μ_o has the exact value of $4\pi \times 10^{-7}$ Wb/m·A (= T·m/A). There is one more magnetic unit that needs mentioning because it is still in the catalogs that offer magnets for sale. The unit is called the gauss. For these purposes 10^4 gauss = 1 T = 1 Wb/m^2. The conversion is correct as long as the measurements are done in air or in vacuum. The conversion is not correct inside magnetic materials. A second specification found in catalogs is the weight the magnet can lift, sometimes given in pounds, sometimes in kilograms.

6.6.3 A small magnet, like a compass needle, is located in the center of a long solenoid where the magnetic field can be considered to be uniform.

The solenoid is characterized by 25 windings per cm length of solenoid and has a diameter of 8.3 cm. At time of the experiment the current flowing in the solenoid is 1.44 A. The magnetic properties of the compass needle are specified by two equal and opposite poles of magnitude 8.88×10^{-5} Wb, separated by 4.22 cm.

a. Calculate the magnetic field inside the long solenoid.

b. The compass needle lies in an arbitrary orientation near the middle of the solenoid. Calculate the force, magnitude and direction, on each pole of the compass. What is the net force on the compass?

c. For three orientations of the compass needle, parallel to the axis, perpendicular to the axis and at 45° to the axis, calculate the torque on the compass needle.

6.7 ALTERNATING CURRENTS

Worldwide almost all electrical power distribution to homes uses alternating (ac) voltage. In North America the norm is a 120 V supply at 60 Hz, in Europe it is mostly 240 V at 50 Hz. The reasons for choosing a given voltage and frequency, and ac versus dc are not obvious. In fact, historically, there was a bitter fight between Edison and the forerunner of General Electric Company insisting on the use of dc and Tesla and Westinghouse pushing for the use of ac in the earliest days of electrical power companies. The ac side eventually won out, but dc into the homes still existed as late as 1950 in Germany. For subways, trolley buses, and trams though the preferred supply is dc at 600 V.

The main advantage of ac over dc is the ease and efficiency with which voltage levels can be changed. The device used is the transformer. Its efficiency is more than 95%. The action of the transformer is analogous to the action of the simple machine of elementary mechanics. In a simple machine mechanical advantage is used to move an object a short distance with a large force by moving a lever arm a large distance while applying a small force. The work done at the lever arm and the work done at the object are the same within the limits of frictional losses. The transformer takes high voltage ac at low current and transforms it to low voltage ac at high current in such a way that, within the limit of dissipative losses, the power in ($I_{in}V_{in}$) is equal to the power out ($I_{out}V_{out}$). The reverse—changing low voltage to high voltage—is equally easily implemented, again subject to small dissipative losses. Changes to dc voltage levels require more complex equipment and additional losses.

A vital application of the transformer is in long distance power transmission. One source of electrical power is a hydroelectric installation located far from consumers. The power must often be transmitted over distances of 1000 km or more along transmission lines which unavoidably introduce resistive losses. Since losses depend on I^2R, there is an advantage in keeping I as small as possible by using transmission voltages that are as large as possible. For example, Quebec Hydro uses 735 kV lines to transmit power from the James Bay generators to the industrial areas of Quebec and the New England states.

6.7.1 This problem leads you to an estimate of the mass of copper that is needed to bring the 2 000 MW of power generated at the first dam at James Bay to Montreal over a 735 kV transmission line and a distance of about 750 km. The added condition is that the resistive power dissipation in the power line be limited to 10% of the generated power. While for technical purposes the current is distributed over three conducting wires, this calculation simplifies the problem by assuming that just two long, thick copper cables to carry the current.

a. Determine the maximum permissible resistance of the cables bringing the current from James Bay to Montreal and back. What is the maximum current that can flow through the cables under those conditions?

b. Determine the diameter of the cables, and from that, their volume and the mass of copper required. The cables will actually include steel in addition to the copper to provide mechanical strength. The electrical conductivity of the steel is small compared to that of copper. Its relative volume will also be small.

c. At the generating station the power is produced at 25 kV and then stepped up to 735 kV for transmission. What is the maximum current that can be pushed through the cables at 25 kV? What then would be the power loss in the transmission to Montreal?

d. What mass of copper would have been required to build a 25 kV transmission line with 10% power loss in the power line from James Bay to Montreal?

COMMENT: As the above calculation shows, there is a great advantage to high voltage transmission. There is a disadvantage of ac over dc. Alternating voltages radiate energy at the transmission frequency. This can cause an audible hum in radios and appliances near the transmission line. There is also a measurable radiative power loss increasing with line voltage and frequency. For some transmission lines the high voltage ac is rectified to become high voltage dc, then transmitted to the location of the users, reconverted there to ac, and finally stepped down to less lethal voltage levels. Manitoba Hydro does this for their Nelson River lines. The dc lines also have the advantage of requiring only two wire cables instead of three. Research continues on the most energy efficient and least intrusive ways to transport electrical energy over large distances. Finally, it is important to compare the energy it takes to transport electrical energy as compared to oil. See the results of the calculations for Problem 3.4.3 for this.

The introduction to Problem 6.7.1 implies that the technology exists to transform ac to dc and vice versa. The following problems deal with one of the problems associated with the technology of rectification, i.e. turning sinusoidal ac to smooth, constant voltage dc. This transformation is particularly important in radios and hi-fi equipment. Internal power supplies must be kept free of 60 Hz signals and its overtones to assure that none of these frequencies find their way into the audio signals in the speakers where they can cause the annoying hum.

There are devices on the market called diodes or rectifiers that can efficiently conduct electricity in one (forward) direction but allow only minute currents to flow through them in the opposite (reverse) direction. For voltage and current levels required in the home, these diodes are made from silicon and are, among other places, embedded in computer chips. For high power levels and high voltages special large vacuum tubes are required. The current-voltage behavior of one such diode was the subject of Problem 6.5.3.

The diagram below shows the action of an assembly of idealized diodes, called a full wave rectifier, on a pure sinusoidal voltage supply.

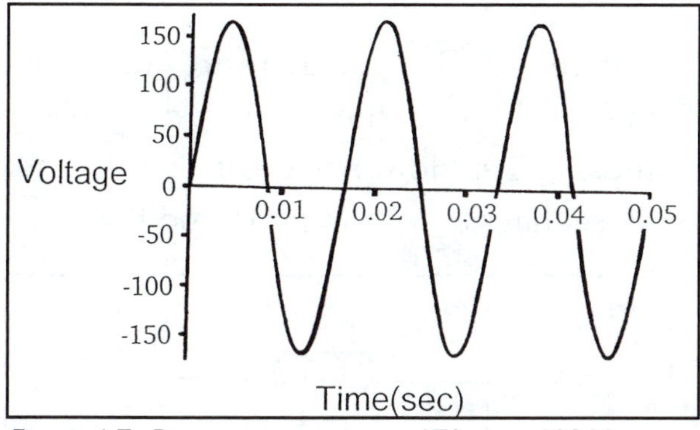

Figure 6.7a Pure ac power input, 170 max, 120 V_{rms}

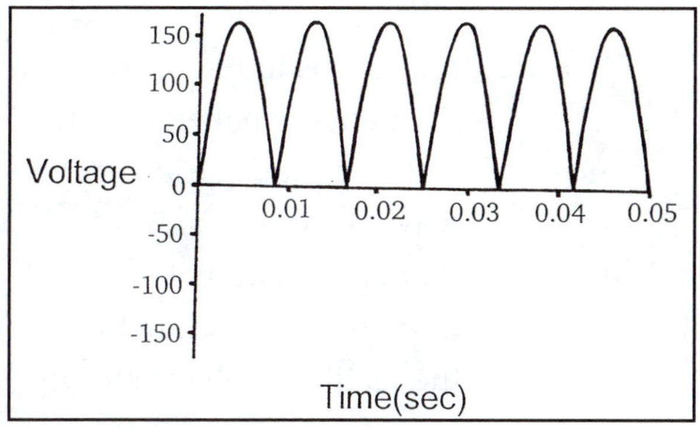

Figure 6.7b AC power rectification, 170 max

The smooth sine wave pattern has been changed to a series of identical humps. Mathematically expressed, the change has been from $V = V_o \sin(\omega t)$ to $V = V_o \cdot |\sin(\omega t)| \cdot$. Whereas $V = V_o \sin(\omega t)$ has an average value of zero, and therefore has no dc component, the rectified signal $V = V_o \cdot |\sin(\omega t)| \cdot$, being positive at all times, has a non-zero average value and consequently a dc component. The next diagram shows how the rectified voltage can be divided into a combination of dc voltage and an oddly shaped ac component.

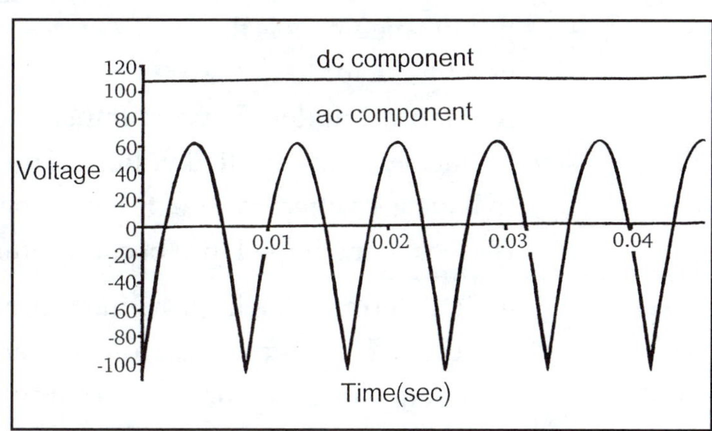

Figure 6.8 Rectified ac acts as a dc component of 108 V plus odd shaped ac component

The mathematical technique called Fourier Analysis provides a series expansion of $V = V_o |\sin(\omega t)|$ which identifies the amplitudes of the dc component and the many frequencies that make up the remaining ac component. The expression is

$$V = V_o \left| \sin\left(\omega t\right) \right| = \frac{2V_o}{\pi} - \frac{4V_o}{\pi}\left(\frac{\cos\left(2\left(\omega t\right)\right)}{(1)(3)} + \frac{\cos\left(4\left(\omega t\right)\right)}{(3)(5)} + \frac{\cos\left(6\left(\omega t\right)\right)}{(5)(7)} + \ldots \right).$$

Although no diode will behave as a mathematically perfect rectifier, the above expression provides a useful guide about what to expect. The dc component is $2V_o/\pi$. The lowest frequency that is generated is not the 60 Hz of the original ac signal, but 120 Hz, followed by 240 Hz, 360 Hz, and other multiples of 120 Hz.

Remember that when referring to the household electric supply, the important consideration is power. While the voltage oscillates between some peak positive and negative value V_o, the power, over one or more complete cycles, is given by the average value of $V_{rms}{}^2/R$. This value is quite different from $V_o{}^2/R$. The stated 120 V of the household electric supply is in fact V_{rms}, while $V_o = \sqrt{2}\, V_{rms} = 170$ V.

6.7.2 The task is to reduce the hum in a stereo set. Because there are large low frequency speakers involved, the power input to the set must be at least 100 W dc and the ac component must be kept low. An idealized full wave rectifier is assumed in the sense that $V = V_o \,|\sin(\omega t)|$ and that there is no internal impedance associated with the rectified voltage supply.

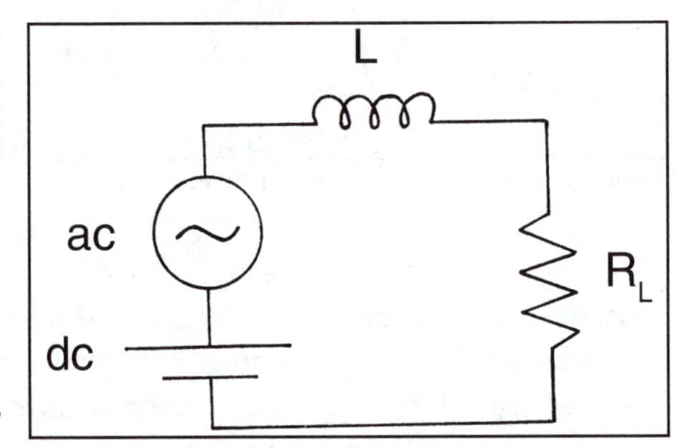

Figure 6.9

Inductors are one way of filtering out the ac component from a mixed ac and dc supply like a rectifier.

The principle of operation for this arrangement is that a good inductor has little resistance, but does have an impedance that increases with the frequency of the applied voltage. Thus the inductor can be considered to have zero effect for the dc component of the supply. At any ac frequency the impedance becomes $\{(\omega L)^2 + R_L{}^2\}^{1/2}$.

a. To determine the improvement provided by the inductor, it is necessary to establish before and after conditions. This is done by first determining the current through the load R without the inductor *L* in place. The peak voltage from the rectifier is 170 V. What is the dc voltage level produced by the rectifier? Next, determine the peak voltage and the rms voltage components of the ac output of the rectifier at 60 Hz, 120 Hz, 240 Hz, and 360 Hz.

b. For the stated required dc power consumption, calculate the load resistance R_L.

c. Still without the inductor in place, determine the power going into the load R_L at dc and the other frequencies listed in *part a*.

d. An inductor $L = 1.5$ H is now added to the circuit in series with the load. Determine the power going into the load R_L at dc and the other listed frequencies.

6.7.3 As in Problem 6.7.2, the task is to reduce the hum in a stereo set. An idealized full wave rectifier is assumed. For high current applications the inductor, as in Problem 6.7.2, is the best method. When the current and power requirements are low, the standard method to reduce ac interference from a mixed ac/dc source is to use a capacitor in parallel with the load as shown in Figure 6.10. The load might be part of the power supply of the stereo set. In some critical filtering applications a parallel capacitor is used to supplement the series inductor.

The idea behind this arrangement is that dc current cannot pass through a capacitor and therefore can only go through the load R_L. On the other hand, ac current can pass through both capacitor whose impedance is $1/\omega C$ and the load whose impedance is R_L. Therefore the circuit has, for ac, two impedances in parallel and the current will divide between the two branches in inverse proportion to their impedances. When $1/\omega C$ is much smaller than R_L, then most of the ac current from the rectified supply goes through the capacitor. There are additional complications that crop up which are illustrated in *parts e* and *f*.

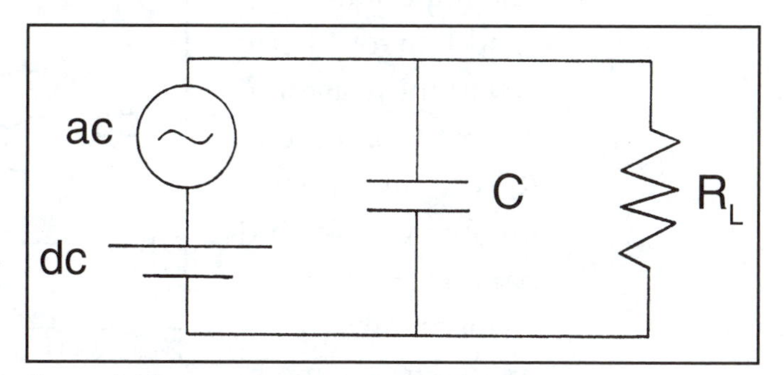

Figure 6.10

If you have not already done problem 6.7.2, do *part a* below. Otherwise skip to *parts b, c,* and *d*.

a. First examine the current through the load R_L without the filter C in place. The peak voltage from the rectifier is 170 V. What is the dc voltage level produced by the rectifier? Next determine the peak voltage and the rms voltage components of the ac output of the rectifier at 60 Hz, 120 Hz, 240 Hz , and 360 Hz.

b. The required dc power consumption is 1.33 W. Calculate the load resistance R_L.

c. Still without the capacitor in place, determine the power going into the load R_L at the three lowest remaining frequencies.

d. The capacitor is now added to the circuit. The capacitance of C is to be such that 9/10 of the 120 Hz ac component flows through the capacitor, and only 1/10 goes through R_L. Determine the value of C to be used.

e. With this appropriate capacitor in place, determine the power going into the load R_L at dc and at the three lowest remaining frequencies. How does that compare with the results from part c?

Any real ac or dc power supply has an internal impedance which limits the current that can be drawn. The more realistic circuit is shown in Figure 6.11, which now includes *R*, the internal resistance of the power supply.

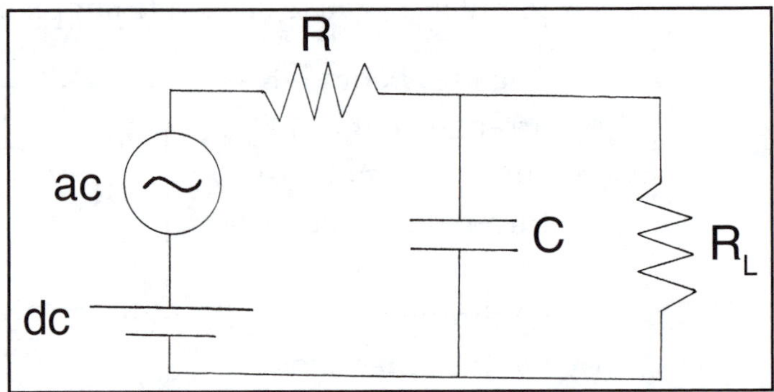

Figure 6.11

The important condition for filtering is that the resistor *R* be small compared to R_L. The calculations involving series and parallel ac circuit elements are unfortunately messy. Until such time as precise evaluations are needed, useful information can be obtained from an approximation. Take the approach that R_L is so large that all the current flows through *R* and *C* and use this assumption to calculate the voltage drop across R_L and *C* caused by *R* at the frequencies of interest.

f. For the calculations that follow, let $R = 880\ \Omega$. The value of *C* is as obtained in *part d*. Calculate the impedance at 120, 240, and 360 Hz for *R* and *C* connected in series, ignoring the presence of R_L.

g. Based on the impedances calculated in *part f* and the ac voltages for the three frequencies as in *part a*, calculate the rms value of the current flowing through the *R C* circuit and the resulting rms voltage drop across *R*.

h. It is now feasible to estimate the ac power dissipated in the load R_L. Since *C* was required to have an ac impedance much smaller that R_L, the peak voltage drop across R_L will be determined by $1/\omega C$ and *R*. At the three frequencies under consideration the voltage across R_L is the voltage supplied as calculated in part a minus the voltage drop across *R* as calculated in *part g*. Determine the voltage drop across R_L and from that the ac power dissipated in R_L at those frequencies.

i. The resistor *R* will also have had an influence on the dc power dissipation in R_L. The original design of R_L was to have a power dissipation of 1.33 W as in *part b*. With *R* in place, determine the power dissipation in R_L.

7 GEOMETRICAL OPTICS

There are some similarities and some major differences between sound and light. For the observer the most obvious difference is the speed with which each travels. Light is virtually instantaneous; for sound there are noticeable delays. Light travels fastest in a vacuum. Sound cannot penetrate a vacuum and indeed travels most rapidly through the most rigid materials. The speed of sound also has a strong temperature dependence. Both light and sound are basically wave phenomena, each with a well-established frequency range. However, each can act as if it consists of individual particles—the photon for light (dealt with in the next chapter) and the lesser known phonon as the particle for sound waves. The phonon is required for the microscopic understanding of the properties of solids, like specific heat and superconductivity.

The science, engineering, and art of illumination is complex. The "standard" eye has quite different sensitivities for different colors and is really an experimental composite based on the measurements of the eyes of many people, with individual idiosyncrasies such as color-blindness averaged out. Instruments used for light level measurements have to be equipped with color filters to desensitize them to those parts of the spectrum (the ultraviolet and the infrared) to which the basic detector may respond, but to which the human eye is insensitive for physiological reasons.

7.1 RAYS

The next set of problems explores some facts relating to illumination and image formation using primarily the $1/r^2$ law and simple geometry.

7.1.1 Shadow casting can be fun on a winter evening, particularly with small children around. Kate enjoys watching the strange shadow shapes her hands create. Some physics and simple geometry are involved to explain why the images can be either sharp or fuzzy, large, or small. A negligibly small, bright light source radiates uniformly over a circular area confined within a solid angle of 1 steradian. A 1.2 m by 1.2 m screen is placed 85 cm from the source of light.

a. If Kate places her finger, which is 5.5 cm long and 1.1 cm wide, against the screen, then how large will the shadow be?

b. If she holds her finger 37 cm away from the screen, then how large will the shadow be?

c. CHALLENGE: What area of the screen is actually illuminated? How close to the light source can she hold her finger and still be able to see the shadow in its entirety? How large will the shadow be in this arrangement?

7.1.2 Linus wants to shadow cast figures on a white wall. His light source is a 100 W incandescent light bulb which can be thought of as a bright sphere, 8.5 cm in diameter. With this arrangement a small object held close to the lamp will barely form a shadow on the wall. The same object held close to the wall will cast a sharp shadow.

The following numerical example will explore the causes. A dollar coin, 26 mm in diameter, hanging from a fine thread between the lamp described above and the wall, casts a shadow. The wall is 92 cm from the center of the lamp.

a. Make two sketches of a set of light rays that originate from the upper and lower edges of the lamp and pass along the upper and lower edges of the coin. One sketch is to depict the situation when the coin is close to the source, the second sketch when the coin is close to the wall. In regions at the wall where there is overlap between rays that reach the wall from above and below the coin, there can only be partial shade. By the same reasoning, the shadow cannot be as sharp as the edge of the coin.

NOTE: In astronomy, in particular during eclipses, one talks about the umbra and penumbra. The umbra is the region of complete shade, for example when during a solar eclipse no light from the sun reaches that spot. The penumbra is a region where some but not all of the light from the sun is blocked.

b. What is the maximum distance between the coin and the wall such that the light is completely blocked from at least one spot on the wall?

c. At what distance from the wall must the coin be located to keep the edges of the shadow sharp within 0.5 mm?

7.1.3. There are strong similarities between X-rays and light rays. The X-ray source is surrounded by thick lead shielding and the rays can only escape in the direction of the patient through a small opening in the shielding. Dense tissue, particularly bones, absorb the X-rays more than other tissues. They form shadows of varying density at the screen or film area. X-rays cannot be seen by the human eye, but fluorescent screens or special photographic films can make the shadow patterns visible. Neither mirrors nor lenses can redirect X-rays, so all medical and dental X-ray diagnostics are based on pinhole camera principles.

a. As a universal principle in taking X-rays, the screen or film is placed as close to the body as possible while the source of the X-rays can be a meter or more away. Why?

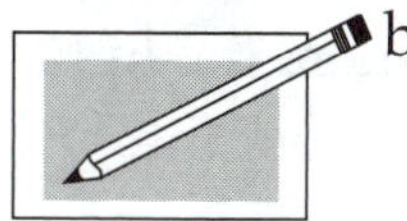

b. The X-ray source is small, but it is not a geometrical point. The object to be imaged—a tooth—is 5 cm from the source, and the screen is 45 cm farther away. By how much would the image of the object be magnified? What is the drawback associated with this magnification?

A pinhole camera is an oversimplified version of all real cameras. Its principles of operation have a validity far beyond the device itself or even cameras in general and are the subject of the following problems. The underlying assumption is that light spreads in all directions from any point on a bright object—the source (of light). Only a small fraction of the light from the source will go in the direction of the pinhole opening of the pinhole camera to land on the screen where it helps to form the image. The sharpness of the image on the screen depends on the distances used and the size of the opening. Aspects of that sharpness can be calculated using simple geometry. Any real camera—still, video, or movie—is based on the same principles but has better light collection mechanisms. These will be dealt with in later sections.

7.1.4 Imagine a uniformly bright spot of light, such as you might get from a TV screen or computer monitor. For the purposes of this problem the spot must be smaller than the complete screen. Therefore the screen is covered with opaque paint, with the exception of a circle 5.0 cm in diameter in the center. That is the "source." A large opaque sheet with a circular hole of 2.0 mm diameter is placed 20 cm in front of the source. That is the pinhole. A white screen is placed 50 cm from the source. Some of the light from the source will go through the circular hole in the black sheet and fall on the screen. That is the "image." Assume that you are working in an otherwise dark room.

a. Draw a diagram of the source, the pinhole, and the screen.

b. Draw a straight line from the center of the source through the center of the pinhole to the white screen. This shows one possible light ray. Draw two more lines from the center of the light source to the screen, this time going through the pinhole along its uppermost and lowermost edges. Light from the center of the source will fall upon those positions of the screen also.

c. Calculate the diameter (in mm) of the area on the white screen which will receive light from the center of the source.

7.1.5 To determine the brightness of the image on the screen under the same conditions as in the previous problem, it is necessary to change perspective and begin from a point on the screen.

a. Starting from a position near the center of the image, draw a series of lines from the screen back to the light source. Draw the lines through the center and along the edges of the pinhole to show where the light at the center of the image could have originated. What area of the source contributes light to the position you chose near the center of the image?

b. Reduce the diameter of the pinhole to one-half its previous value, from 2.0 mm to 1.0 mm. How large an area of the source now contributes light to that center position on the screen?

c. Because the source was assumed to be uniform in intensity and because the angles are small and the distances large, it is safe to assume that all the relevant points from the source which illuminate each given point on the screen will contribute equally to the brightness at the screen. If under the assumptions of *part a* the light intensity at the screen were to be I_o, then what will be the intensity at the screen under the conditions of *part b*?

d. With the original 2.0 mm diameter pinhole the center-most position of the source could spread its light over a certain area of the screen. With the pinhole reduced to half its previous diameter, how large an area on the screen will that same position of the source now illuminate?

To summarize the ideas dealt with in the previous two problems about the pinhole camera: a change in the size of the pinhole has two consequences. Reduce the size of the pinhole and the image becomes sharper, but at the expense of the brightness of the image. In later sections it will be shown that a mirror or a lens substituted for the pinhole can improve both sharpness and brightness, but again at a cost.

7.1.6 The relative distances between source, pinhole, and screen not only influence the sharpness and brightness of the image, they also determine the size of the image. Try it with numbers.

The hot filament in an automobile headlamp is 3.3 mm long and 1.6 mm wide. It is positioned 5.2 m from a "pinhole" opening 2.4 mm in diameter in an opaque sheet. A screen, located in line with the lamp and the pinhole, is 55 cm farther away.

a. Calculate the area on the screen which will be illuminated by a small (0.50 mm by 0.50 mm) region of the center of the filament.

b. Calculate the outer dimensions of the illuminated portion on the screen from the entire filament.

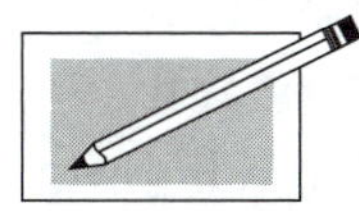

c. The filament is then moved twice as far from the opening. The screen stays where it is. Repeat the calculations required for *parts a* and *b*. What conclusions can you draw with regard to changes in brightness and sharpness of the image?

7.1.7 The moon is a large, bright object far away and could serve as a source for a pinhole camera image. The important astronomical details are that the radius of the moon is 1.738×10^6 m and that the average distance from the center of the Earth to the center of the moon is 3.844×10^8 m. It is also useful to know that the average radius of the Earth is 6.368×10^6 m.

a. As seen from the surface of the Earth, calculate the angle between two rays of light coming to the eye from the moon if one ray comes from the top edge of the moon and the other ray comes from the bottom edge of the moon.

b. Instead of going to the eye the two rays continue in a straight line through a pinhole and then to a screen 35.0 cm behind the pinhole. Calculate the size of the image of the moon. Recalculate the size of the image if the screen were 70.0 cm behind the pinhole. How would the brightness of these two images compare?

c. For a pinhole 0.80 mm in diameter calculate the area over which the light from a point on the moon will be spread on the screen placed 35.0 cm from the pinhole. Repeat the calculation based on a distance of 70.0 cm from the pinhole to the screen.

d. For each case considered in *parts b* and *c* calculate the ratio of the area of the total image of the moon and the area illuminated by the light from just one point on the moon.

7.1.8 The sun is a distant object that is an ideal object for a pinhole camera, particularly because it is so bright. Observing solar eclipses becomes particularly simple. The pinhole camera need only be a large opaque sheet with a small hole held near a wall. The wall becomes the screen where the eclipse can safely be observed

The diameter of the sun is 1.392×10^9 m. The average distance from the center of the Earth to the center of the sun is 1.496×10^{11} m.

a. As seen from the Earth, calculate the angle between a ray coming from the top edge of the sun and a ray coming from the bottom edge of the sun.

b. The two rays from the upper and the lower edges of the sun pass through a pinhole to a screen which is 35.0 cm behind the pinhole. Calculate the area of the image of the sun on the screen. Repeat the calculation for the arrangement in which the screen is 70.0 cm behind the pinhole. Compare the brightness of the two images.

c. For a pinhole 0.75 mm in diameter calculate the area over which the light from a point on the sun will be spread on the screen placed 35.0 cm from the pinhole. Repeat the calculation when the screen is 70.0 cm from the pinhole.

d. For each case considered in *parts b* and *c*, calculate the ratio of the area of the total image of the sun and the area illuminated by the light from just one point on the sun.

7.1.9 Occasionally—during a solar or lunar eclipse—the sun and the moon can be in a straight line with an observer on the Earth.

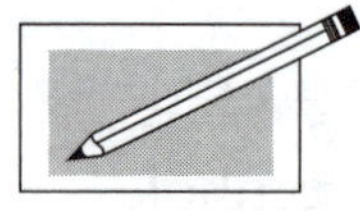

a. Based on the calculations from Problems 7.1.7 and 7.1.8, is it possible for the sun to completely block the light from the moon? Is it possible for the moon to completely block the light from the sun to that Earthbound observer? Explain for both situations why or why not.

b. There are complications in the orbits of the Earth and the moon which make eclipses variable in character. For example, the center-to-center distances between the Earth and the moon and between the Earth and the sun change appreciably because orbits are elliptical rather than circular. The greatest distance, center to center, between the sun and the Earth is 1.521×10^{11} m each July. The closest distance from the center of the Earth to the center of the moon is 3.56×10^{8} m. How will these facts change the relative sizes of the images of the sun and the moon on the screen of a pinhole camera, and therefore the possibility of a complete solar eclipse?

7.1.10 Human depth perception is partly explained by binocular vision. Two eyes see the same object from slightly different directions, and the brain, processing this information, judges the distance. The following problem takes one of the possible approaches the brain may use.

Two eyes in a face are 6.5 cm apart, center to center. The letter "A" is 60 cm directly in front of the right eye. The left eye initially stares straight ahead (Figure 7.1).

a. How many degrees must the left eye turn to face the same letter "A"? Repeat the calculation with the letter "A" at 25 cm and then at 10 m in front of the right eye. This change in angle can be one possible input for depth perception.

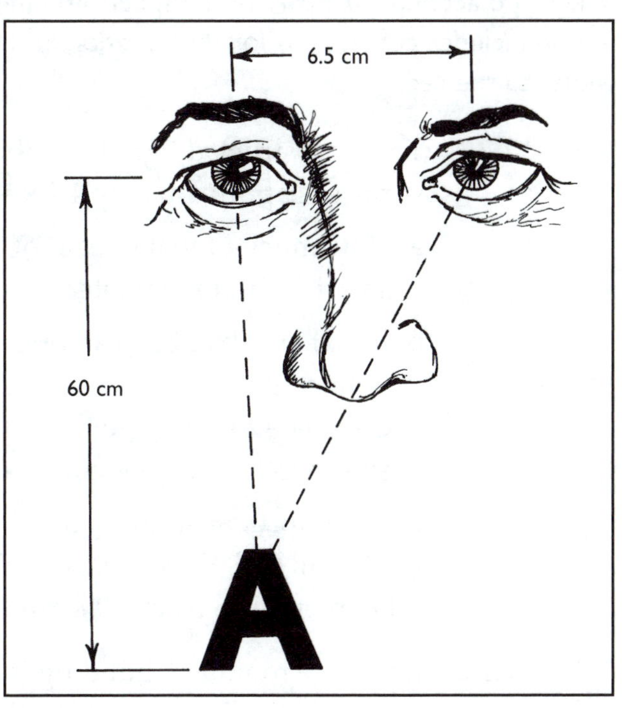

Figure 7.1

b. Continue the calculations of the difference in the angle between right and left eye when the distances become 50 m, 500 m, and 5000 m. As the letter "A" becomes too small to see, it is necessary to think in terms of a larger object like a telephone pole or a grain elevator.

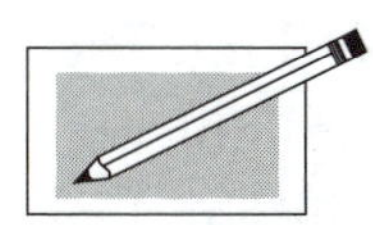

c. How small an angle do you think the human physiology can "measure"? Explain how you think humans judge the greater distances.

7.2 LIGHT LEVELS

Light levels vary over a wide range. While the human eye has preferences for a limited range of light values for reading or working, it can tolerate a wider range of intensities. Moonlight at midnight is fine for romance whereas the midday sun in the spring on the ski slopes can cause snow blindness.

The following problems will lead you through some calculations on the brightness of various light sources and what is required for reasonable light levels. For the purposes of the following problems an analogy is useful. Light can be thought of as a continuous stream of fluid, such as water. A water stream has a source which might be a lake or a spring, while the source for light might be the sun or a light bulb. Water is absorbed in dry ground, while light is absorbed by colored surfaces, particularly black ones.

Illumination engineers use many different units of measurements to describe the situations they have to deal with. The following problems are restricted to only one basic unit, the lumen (lm). In terms of our analogy, the number of lumens on an area is similar to the number of liters per second of water flowing to or through an area. The lumen not only takes into account that part of the spectrum emitted by the source which the eye can detect, it also includes corrections for the variation in the sensitivity of the "standard eye" from the violet to the red.

7.2.1 The unobstructed light from the sun that can fall on 1.0 m^2 at the Earth's surface is approximately 4.4×10^5 lumens.

a. How much of that stream of unobstructed light would fall on a round area 1.2 mm in diameter?

b. If that area had been an opening in an opaque sheet (a pinhole), the light could continue on to a screen 35 cm behind the opening. Calculate the average illumination level of the sun's image on that screen in lumens per cm^2.

c. A comfortable illumination level for reading is considered to be about 1000 lm/m^2. How large should the opening in the black sheet of *part b* be in order to reach 1000 lm/m^2 at the image location?

7.2.2 An entire room in a building is painted with a perfect white paint, including the floor and the ceiling. This means that all the light entering the room from the window is uniformly distributed throughout the room by many reflections from the walls, the floor, and the ceiling. The room is rectangular, 2.5 m high and 3.2 m wide and deep. Sunlight enters the room through a window at 2.5×10^5 lm/m^2. (Absolutely clear skies with the sun shining perpendicular to a window is a rare occurrence.)

a. What is the total surface area of the room including the window?

b. On this sunny day how large would the window have to be in order to achieve the recommended illumination level of 1000 lm/m^2?

7.2.3 A room without windows has a ceiling 2.5 m high and a floor that is 3.5 m by 4.2 m. All walls and ceiling are painted with a perfect white paint to distribute the light uniformly. One ordinary 100 watt (W) frosted incandescent lamp has a total light output of 1700 lumens. How many of these lamps would be required to achieve the recommended illumination level of 1000 lm/m^2?

At home and at work you pay for energy used in terms of watts consumed over so many hours. Good working conditions require adequate light levels. It is therefore important to be able to compare light sources in terms of lumens (light flow out) per watt (energy flow in). The real cost in providing light also includes how much you pay for the particular fixture and how often it will need replacement. Sunlight as a light source may seem free, but the architect for a commercial building must include in his calculations the cost of windows versus plain walls and the cost of the air conditioning that might be required when the heat from the sun enters a window along with the light.

7.2.4 The energy use and light output of the wide variety of manufactured light sources has been measured and can be found in handbooks used by illumination engineers. There are three types of lamps commonly used in homes and offices. The most common are the frosted bulb incandescent lamps. Then there are fluorescent lamps, some of which (compact fluorescents) have recently become available in a form which can be used in ordinary incandescent lamp fixtures. The third type of lamp of interest is the quartz halide lamp. They are most often used in automobile headlamps, as small spot lights, and in slide projectors.

A standard 100 W incandescent lamp produces 1700 lumens. A 30 W "cool white" fluorescent lamp produces 2400 lumens, while a 50 W quartz halide lamp is rated at 1600 lumens.

a. For each of these lamps calculate the "efficacy," defined as light output per unit energy input (lumens/watt).

b. The illumination requirements for office space are 1400 lm/m^2. A particular office has a 3.6 m by 12 m floor area and a ceiling height of 3.2 m. The office space can be illuminated by any of the three types of lamps mentioned above. For each of the three types of lamps, calculate how many would be needed and what the energy consumption would be to achieve the required light level. (Assume that the light is uniformly distributed thanks to careful placing of the lamps and highly reflecting walls, floor, and ceiling.)

7.2.5 For outdoor use, such as the illumination of highway intersections, the tax burden is more important than good color rendition. Fluorescent lamps are too sensitive to temperature, so the choice is usually between the orange-yellow, high-pressure sodium lamps and bluish-white mercury arc lamps. They are efficient in converting electrical energy to light but distort the natural colors.

Typical lamps used for outdoor illumination are 400 W high pressure sodium lamps at 125 lumens per watt and 400 W mercury lamps at 52 lumens per watt. The lamps are mounted 11 m above the ground and are assumed to have perfect reflectors above them to direct the light to the street instead of lighting up the sky.

a. According to highway engineers it is neither necessary nor advisable to achieve on roads the light levels that are required for comfortable reading. An average illumination level of 9 lm/m^2 is considered adequate for highway use. A minor highway intersection is shaped like a cross with each of the 4 arms 30 m wide and 425 m long. Calculate how many lamps of either type are needed to get the required light level.

b. Calculate the amount of electrical power in watts is required to light the complete intersections for each type of lamp.

c. The use of 1000 watts for a period of 1 hour is called a kilowatt-hour (kWh) and costs about 4 cents (wholesale!). How much will it cost the Highways Department to light up the intersection described in *part a* for one year? Assume that, on average, the lights will be on for a 12-hour period each night.

7.2.6 Lasers are the most intriguing light sources these days. From the point of efficacy, i.e. the visible light output for a given electrical energy input (see Problem 7.2.4a), they are abominably poor. The common red (632.816 nm wavelength) helium-neon gas laser has a light energy output of 2 mW, which translates to 0.35 lumens. The electrical power input that is needed to get the 2 mW output is 13 W.

a. Calculate the efficacy of this device.

b. Discuss the uses of such a device in spite of the low efficacy.

7.3 REFLECTION

Mirrors, both plane and curved, are common commodities. They are used in two distinct ways. In one application the mirror is used to redirect light, often without regard for the perfection of the surface involved. The reflector behind the lamp in a flashlight is an example of this application. The second application is to enable a person to see around corners or behind himself. The rearview mirror in a car is an example of this second application. In this case the light from a car behind you has been redirected to your eye by the mirror. Therefore, when you look at an image in a good mirror, you perceive the object as being located somewhere on the other side of the mirror and the mirror itself becomes invisible.

The expressions "it's all done with mirrors" and "smoke and mirrors" hint at the fact that the use of mirrors can fool the eye. Flat mirrors act in ways similar to windows or openings. When you look at yourself in a mirror you see yourself and other objects behind the mirror. It is as if you were looking at someone much like yourself behind a window. An object behind a window or an opening is seen because light from the object travels in a straight line through the window or opening to the eye. An object is perceived to be behind a mirror because the light from the object seems to come straight through the mirror to the observer. The observer (eye) is not always aware that there is a mirror between the object and the observer.

There is a very important difference between a mirror and a window. Even a small rotation of a mirror changes the objects we see. Rotate a window by an equally small amount and all that will change is the effective size of the window because the edges have moved. Given a stationary light source, a window, and a stationary observer, light will reach the observer from the source only if the window is in a straight line between source and observer. Given the same source and observer but a mirror instead of a window, there will be many locations where the mirror can be placed to redirect the light from the source to the observer.

The next few problems explore the mirror-window contrasts.

7.3.1 This problem is similar to the pinhole camera problems in a previous section. The pinhole, however, has been replaced by a larger diameter mirror.

There is a uniformly bright, round light source with a radius of 2.3 mm, similar to the filament of a light bulb. A round flat mirror 2 cm in diameter is placed 43 cm from the light source such that the mirror's front surface makes an angle of 36° with respect to the beam from the light source.

a. How many degrees will the beam deviate from its original direction after leaving the mirror?

b. There is a screen facing the mirror at a distance of 85 cm. How large an area on the screen will be illuminated from any point located on the source? What shape will it have?

c. Calculate the area of the screen which will be illuminated by the reflected light from the source. What shape will it have?

d. The mirror is replaced by an opaque sheet with an opening to change the system to a pinhole camera. The opening is the size and shape of the mirror. Where must the light source be relocated to illuminate the screen in a manner identical to *part b*? Explain what happens to the shape and location of the illuminated area as the opening is rotated about an axis along its diameter. Contrast that to the results of rotating the mirror.

Light rays that are initially going in different directions can be intercepted by additional mirrors and redirected to a common location, thereby increasing the brightness at that location. The advantages and consequences are explored in the following problems.

7.3.2 Figure 7.2 shows a spherical light source, 0.35 mm in diameter, shining uniformly in all directions. A flat mirror (A), 1.0 cm² in area, is placed 56 cm straight to the left of the light source. The mirror reflects the light back toward the source at 165° from its original direction of travel.

a. What fraction of the light output of the source is intercepted and redirected by the mirror?

b. A second flat mirror (B), also 1.0 cm² in size, is placed below and adjacent to the first mirror. It is tilted so that the rays reflected from this mirror intersect the beam from the first mirror at a distance of 34 cm from the first mirror. What fraction of the light from the source does the second mirror redirect?

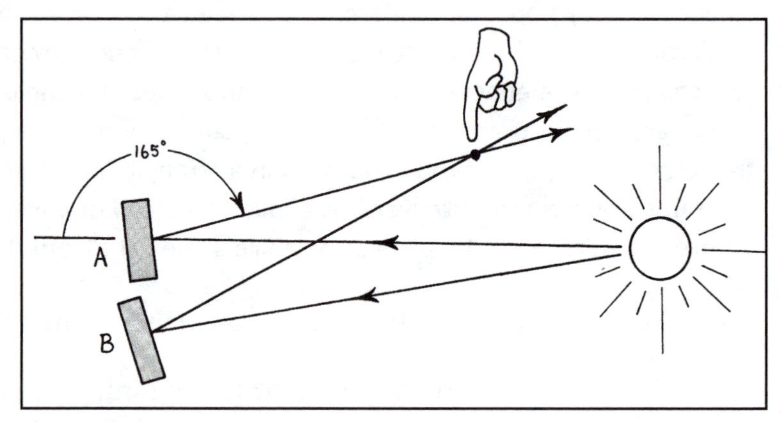

Figure 7.2

c. Determine the area that the intersecting light beams will cover where they cross at 34 cm from the mirrors. How much brighter will that area become? Where can additional mirrors be added to further enhance the brightness at that location?

d. Once you have the two mirrors adjusted so that their beams intersect at the 34 cm mark, how much farther away in the same direction from the mirror will the two beams no longer overlap?

e. It is still possible at this new distance from the mirrors to get the beams back together without moving the mirrors. The light source must be moved instead. Where should the source be placed to get full overlap of the two beams at the new distance from the mirrors determined in *part d*.

7.3.3 A star can be considered to be a point light source at an infinite distance. From an optics point of view this means that the light rays from the star to the Earthbound observer (wherever that observer might be) can be considered to travel parallel to each other. Carefully shaped mirrors in large astronomical observatories are designed to collect as much starlight as possible and concentrate it on a detector. The detector may be the eye of the astronomer, photographic film, or an electronic device. The fainter the star, the more light that has to be gathered and brought to the detector. Figure 7.3 shows how this might be done. The unaided, dark-adapted eye can only intercept as much light as hits the pupil of the eye—an area of about 30 mm^2. It does not matter how far the observer on Earth moves toward or away from the star because the distance from the star to the Earth is so great.

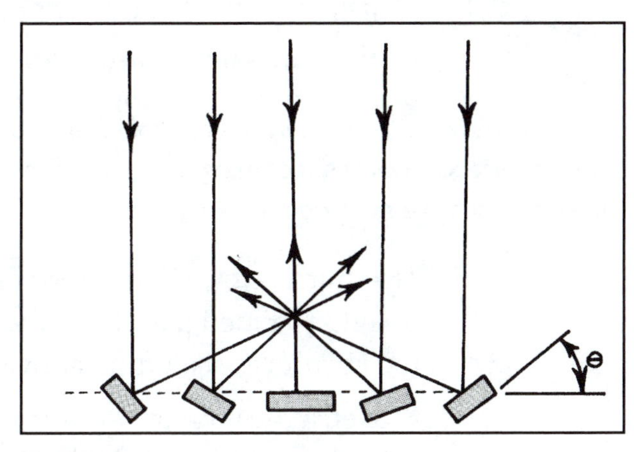

Figure 7.3

a. A flat mirror can be used to reflect the light from a single star back to the eye. If the area of the pupil of the eye is as stated above, then what should be the diameter of the flat mirror so that none of the light needed by the eye is lost and that none of the light is wasted by being directed elsewhere? Does this depend on the distance from mirror to eye?

b. Additional mirrors can be placed along a line on the same plane as the first one, so that each of these additional mirrors directs light to the pupil of the eye located at 1.0 m above the first mirror. Draw a graph of the angle to which a mirror has to be turned as a function of its distance from the first mirror. Repeat, for a few points only, to show how the graph would change if the eye were at 10 m instead of 1.0 m above the first mirror.

c. Because the mirrors are slanted more and more as their distance from the center mirror increases, they will redirect a smaller fraction of the light. Calculate the distance from the center mirror at which each mirror will still reflect more than 90% of the light it receives as compared to the central mirror to the detector (placed at 1.0 m). How large an area could be covered by the small mirrors that fit the 90% criterion? How is this related to the brightness of the star as detected by the observer?

NOTE: In telescopes the individual small flat mirrors are replaced by one large mirror, smoothly curved in the manner you have calculated, with correspondingly large light-gathering power.

7.3.4 A convention has been adopted by ambulance services. The word "AMBULANCE" is consistently written backwards on the front of the vehicle. On all other sides of the vehicle conventional writing is used. Explain the safety reasons for this convention.

Combinations of plane mirrors can make light beams bounce around many times depending on the direction of the incoming beam and the angle between the mirrors. The following special cases are noteworthy.

7.3.5 Two mirrored walls in a room touch each other at an angle of 90°. A ray of light is aimed parallel to the floor and makes an angle θ with one of the mirrors as seen from above in Figure 7.4.

a. Sketch the path of the beam before and after its reflection from the first and from the second mirror. Determine the angle to the first mirror at which the beam leaves the set of mirrors.

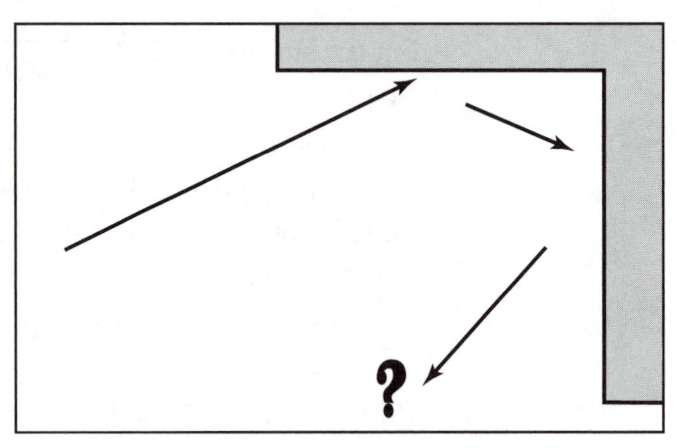

Figure 7.4

b. Looking horizontally into the corner where the two mirrors meet, what can be seen? How does that image differ from the image seen looking perpendicularly into only one of the mirrors?

7.3.6 Three mirrors meet seamlessly at a corner in a room where two walls and the ceiling join. Each is precisely perpendicular to the others. A ray enters the room at an angle θ with the north-mirrored wall and at an angle ϕ with the mirrored ceiling. The ray hits the mirror on the north wall first, then the mirror on the ceiling, and finally, the mirror on the east wall.

a. Sketch the path of the beam before and after its reflection from the first, second, and finally from the third mirror. Determine the angle at which the beam leaves the set of mirrors with respect to the first mirror and the ceiling.

b. A beam of light from an arbitrary direction is aimed at a spot close to the corner where the three mirrors meet. Where will the light go after its multiple reflections?

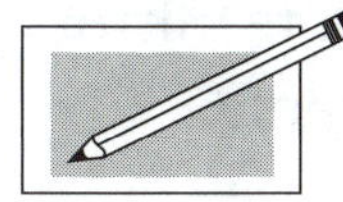

c. The principle of three mutually perpendicular reflectors can be more easily implemented with prisms shaped like the corner of a cube. Such prisms are used for reflectors along highways and on traffic signs. Explain the advantage of such reflectors for that application.

7.3.7 Ryan is facing the large wall mirror in a hair salon admiring his new hair style. The stylist holds a hand-mirror behind his head so he can admire the work. Explain how and why he can see the back of his head. When Ryan looks at himself in a mirror he doesn't see himself as others see him. Why? Will Ryan see the back of his head in the same way as his stylist sees it? If not, how will it differ?

7.3.8 While buying a new jacket Nate is standing in front of a set of three mirrors in the fitting room. Looking straight ahead in the central mirror everything looks just as it does in front of a mirror at home. The other two mirrors are placed perpendicular to the central mirror to its left and to its right sides. Nate can look into any of the mirrors directly or indirectly, making use of multiple reflections.

a. Figure 7.5 shows Nate's location and the three mirrors. In which directions must he look to see an image of himself? What will be the possible locations of his images behind the mirrors?

b. Some of the images will be identical to what he sees in a single mirror. Other images will be different. How will these images differ?

(**HINT:** Think specifically about differences in hand-eye coordination. If possible look in the mirrors of a clothing store.)

Figure 7.5

7.3.9 Problems 7.3.7 and 7.3.8 are about the images from two or more mirrors facing each other at specific directions. This problem is of a more exploratory type. It can be done with pencil and paper, but it may be amusing to do the investigation with the help of two mirrors that can be moved around. If you become interested in the implications, then be prepared to spend hours playing with all the possibilities.

With a single plane mirror it is only possible to get a single well-defined image from a single object. To see the image in the mirror there is a restriction on the location of the observer. If the object is located too far up, down, or sideways, then the image is no longer seen. Two mirrors placed parallel and facing each other, as are used at hair salons, form an almost infinite number of images. Two mirrors placed at exact right angles to each other form, at best, three images of the object. There are also restrictions on the location of the observer to be able to see all three images.

a. Start with an obtuse angle between the prisms and show that there can be as many as four images but that a single position of the observer can only pick up, at most, three of the four images. As part of the solution draw a diagram showing the mirrors, the object, and the locations of the images. Show in what direction the observer must look to see any one of the images. What is the maximum number of times a beam of light can bounce off any of the mirrors?

b. Continue your exploration with an acute angle between the mirrors, first with an angle just slightly less than 90°, then with smaller angles. Explore the number of times a beam can bounce between the mirrors before leaving the system. Then try to find the possible image positions and where they can be seen from.

A newly married couple move into their new home. At their request the bedroom window is made of "one-way glass" to achieve both privacy and good illumination from the outside. In subsequent weeks they find that the neighbors smile at them a bit too knowingly. A careful check shows that the window has been installed the wrong way around.

This story is an urban myth. The science behind the story is that glass can be coated with thin metallic films. The thickness of the film can be carefully controlled to allow a predetermined fraction (for example, 5%) of the energy of the light incident from <u>either</u> side to be transmitted through the glass and film. Most of the rest of the light (almost 95%) will be reflected back toward the source. A fraction of a percent of the incident light is absorbed in the metallic film. The fraction of the light reflected and transmitted does not dependent on which side of the glass is coated.

Other thin films made of transparent materials (called dielectric coatings) can be coated on window materials to make filters to almost any specification. For example, they can make the window opaque for all but a narrow wavelength range.These are called narrow band pass filters and are widely used in optics laboratories. The opposite is also possible. Coatings can be created which are transparent for all but a narrow band of wavelengths, while at the selected wavelength the energy reflected is 99% or more and only 1% or less is transmitted. These dielectric coatings films also have the important property that the incident light is either transmitted or reflected with virtually no losses due to absorption. Devices like helium-neon lasers require these types of filters.

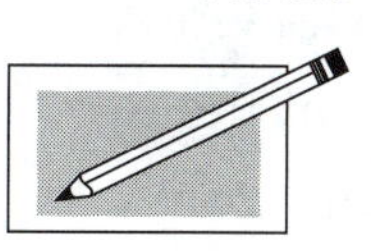

7.3.10 The design of the "one-way mirror" has been explained previously. The task for this problem is to explain why the story related above must be an urban myth. Under what conditions can you see through the "mirror" and under what conditions does it act like an ordinary mirror? Approach the problem from the point of view of two observers, Annabelle on one side of the mirror and Zelda on the other side. Each has two lamps next to her, a bright one and a night light, which they can turn on or off at will.

7.3.11 It is a requirement for the operation of a gas laser, such as the red-emitting helium-neon laser, that the light at the laser wavelength bounces back and forth through the electrically excited gas to pick up power. The requirement can be compared to a diver on a diving board bouncing a few times to gain height before the dive.

A helium-neon laser has been built with two dielectric coated mirrors about 20 cm apart and mounted absolutely parallel to each other. One mirror reflects 98.9% of the light hitting it, the other reflects 96.2%. All other light is transmitted.

a. A certain light intensity (energy) I at the laser wavelength has been injected between the mirrors at a certain moment. This light then bounces back and forth between the mirrors without further energy input. How many return trips will the light take before its intensity drops to $0.10I$?

b. Given the distance between the mirrors, how long will it take before the light intensity drops to $0.10I$? What distance has the remaining light traveled in that time?

c. In the operation of an actual laser the pulse gains strength each time it passes through the gas from the electrical excitation of the gas mixture. Suppose that this gain is 2.0 mW per round trip. How much light will come out of each mirror when a steady state has been reached? Under those conditions what will be the intensity I inside the gas?

7.3.12 Optical devices called range finders depend on partially reflecting and partially transmitting mirrors to permit the operator to measure the distance to an object. The military may need to know the distance to a target, or a photographer may require an accurate measurement of the distance to his subject. The principle of operation is similar to that of binocular vision. Two sets of rays from the object to the instrument are intercepted and directed to one eye of the observer. One set of rays follows a straight path from object to eye; the other leaves the object at a slight angle and then, after reflections from two mirrors as shown in Figure 7.6, rejoins the first set of rays to the eye. The mirror labeled A can be accurately turned to superimpose the two sets of rays at the eye. Mirror B can acts as both a mirror for the light coming from mirror A and as a transparent window for the direct beam from the object.

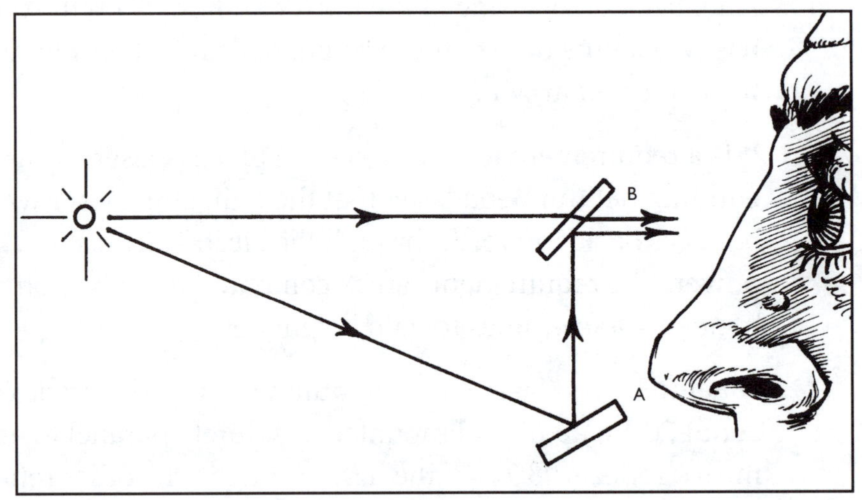

Figure 7.6

a. If the distance between mirrors A and B is 12.5 cm, then over what range of angles should mirror A be turned to get the two beams to coincide at the eye when the object distance varies from 50 cm to 200 m?

b. Plot a graph of object distance vs angle of mirror A.

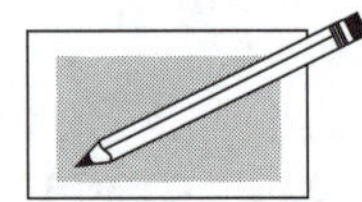

c. Explain how the principles used in this device relate to depth perception in binocular vision. (Refer to Problem 7.1.10)

7.4 REFRACTION

Light has the interesting property of being able to penetrate solid materials. Most substances, particularly if they are thin enough, let a small amount of light pass through but at the expense of destroying the initial direction of the incoming beam. Light passing through a sheet of white paper reduces and diffuses in this way. There also are a limited number of natural and manufactured substances that have the unique and useful property of keeping light rays relatively intact as they pass through. Glass is the most common material of this type.

The law of refraction is called Snell's Law and is written as $n_1 \sin\theta_1 = n_2 \sin\theta_2$. The indices of refraction, n_1 and n_2, are experimentally determined properties which allow us to predict the path a ray of light will take when entering a new substance. The index of refraction varies from substance to substance and also, for each given substance, on the temperature and on the wavelength of the light. Rainbows and the colorful sparkle of faceted crystals are caused by index of refraction changes.

7.4.1 A straight stick, 1.00 m long, is placed halfway into the water making an angle of 47° with the surface of the water. The index of refraction of water is 1.33.

a. Prove that if you were to sight straight along the stick above the water you would not see the end of the stick which is submerged in the water.

b. Where should you look above the water to sight along the submerged end of the stick?

7.4.2 Objects under water never seem to be quite where they appear to be. For example, imagine looking straight down into a tub of water at a toy, as in Figure 7.7. Your eyes are 15 cm above the water level, 5.5 cm apart, each aimed 5° toward the center. The rays from the toy directed to your eyes will have had a change in direction at the air-water interface. The index of refraction of water is 1.333 at 20°C.

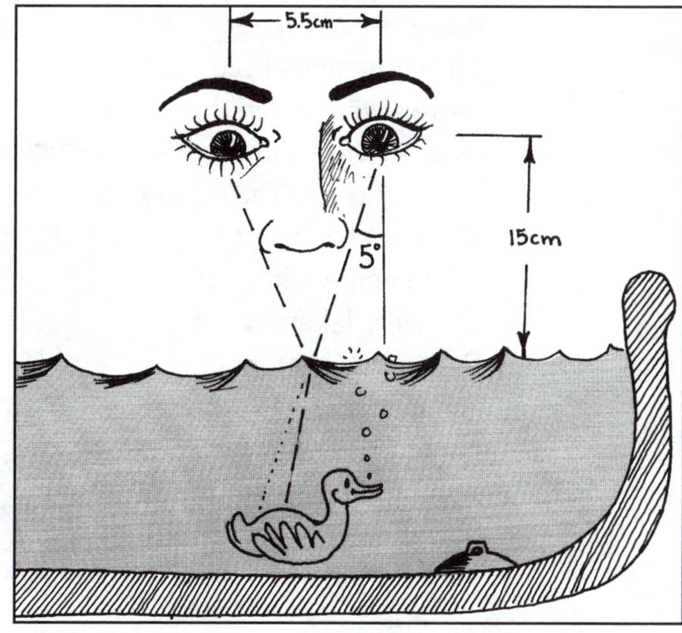

Figure 7.7

a. The rays reaching the eyes leave the water at an angle of 5° to the normal to reach the eye. At what angle to the normal did they travel before meeting the air-water interface?

b. How far below the surface of the water must the object be in order to be seen this way? How far below the eye is that? The result of this calculation gives you the object distance, that is, the true distance from the object to you the observer.

c. The brain interprets the location of the object by ignoring the existence of the air-water interface. Calculate, as the brain apparently does, how far below the eyes an object must be if it is seen only through the air and if the eyes are 5.5 cm apart aimed 5° toward the center. The result of this calculation is the real object distance if there was only air between the object and the eyes. If, because of a change in refractive index, the object is actually elsewhere, then this same calculation gives the "apparent distance" where an image of the object is located.

d. The water level was stated as 15 cm below the eyes. How far below the surface of the water is the image of the toy? Calculate the ratio between the distances calculated in *part b* and in *part d*.

7.4.3 Looking at the bottom of a swimming pool, while walking along its edge, you can clearly see distortions even when the water is completely calm. The distortions become more severe when the eyes are closer to the water surface. The effects become disconcerting when you walk hip deep in a clear lake or pool. Even if the bottom is absolutely level, it looks curved and your feet don't quite seem to touch bottom where you think they should.

Figure 7.8

Figure 7.8 shows the Hunk standing in uniformly 1.20 m deep water. His eyes are 0.60 m above the surface and scan the bottom for shells.

a. If the starfish was lying next to his toes, at what depth would they appear to be?

b. The Hunk is not quite so fortunate. He sees a starfish when he looks downward at the surface of the water at an angle θ with the vertical. Derive expressions for the actual x and y coordinates of the position of the starfish that the Hunk sees at the angle θ. Use the surface of the water just below his eyes as the origin of the coordinate system.

c. Calculate the distance L the light must travel in the water from the starfish to the surface of the water to be seen by the Hunk at the angle θ.

d. The apparent distance the light travels is reduced because any distance in any transparent material appears shortened along the line of sight, in the same fashion as the expression for apparent depth. What is the distance L' the Hunk thinks he has to extend his arm under the water to grab the starfish he sees at the angle θ?

e. Derive expressions for the coordinates x' and y' (use the same coordinate system as in *part b* of the apparent position of the starfish, as seen by the Hunk, as a function of the angle θ and the distance L.

f. On the basis of the expressions derived in previous parts of this problem draw graphs of apparent depth and apparent horizontal position as a function of the angle θ of the starfish.

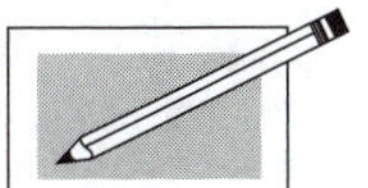

g. Based on the graphs you have drawn in *part f*, explain how walking in a pool or lake will be a disorienting experience.

The laws of reflection and refraction determine the properties of prisms. Prisms are used for two fundamentally different purposes. One use is to split the spectrum into its component colors based on the changes of the index of refraction with wavelength. The second and highly practical use of prisms is in optical instruments where a beam of light has to be redirected.

Binoculars have prisms in them, as do viewfinders in many cameras. Although sets of mirrors could be used instead of prisms, high-quality prisms lose less light in an optical system when the principle of total internal reflection is used. Also, from a manufacturing point of view, it is easier to keep one prism properly aligned than two or more individual mirrors. The following problems explore the uses of prisms of one basic shape. Although the mathematics is simple, the results are sometimes surprising.

Of primary importance for the use of prisms is to determine the position and the orientation of an object as seen through a prism. To set the stage, refer again to the plane mirror. Figure 7.9 shows an object and two of the many rays that leave the object in the direction of the observer. The eye, in trying to see the object, looks in the direction the ray comes from immediately before the light enters the eye. A high-quality mirror is one that the brain ignores as an intermediary entity. If the light after its last reflection comes from the left, then the brain interprets whatever is seen as coming strictly from the left. The observer sees a virtual image, that is, he or she sees something in a position where the light has not been. Figure 7.10 shows the same object but with the mirror slightly rotated. The light rays from the object have been redirected and, as a consequence, the image has shifted significantly.

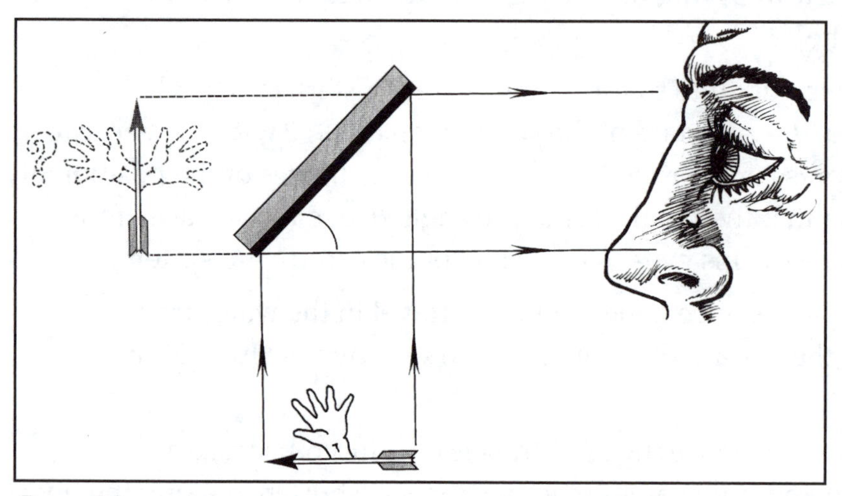

Figure 7.9

Figure 7.10

Prisms come in many different shapes for different applications. Some of these applications will be explored in the following problems. The most commonly used prism is shaped like an isosceles right triangle (45°, 90°, 45° angles). The glass type most often used for high-quality prisms is one labeled BK 7 and has an index of refraction of approximately 1.52. The image seen when looking at an object through such a prism varies significantly with the orientation of the prism. The next four problems ask almost identical questions about the same prism used in different orientations.

7.4.4 Figure 7.11 shows one possible orientation of object, prism, and observer. One of the rays from object to observer is also shown. The prism has the shape of an isosceles right triangle. The index of refraction of the glass is 1.52.

a. For the incoming ray shown, which need not be perpendicular to the bottom face of the prism, calculate the path of the ray as it enters the prism, is reflected at the slanted face, and emerges again on the right side of the prism.

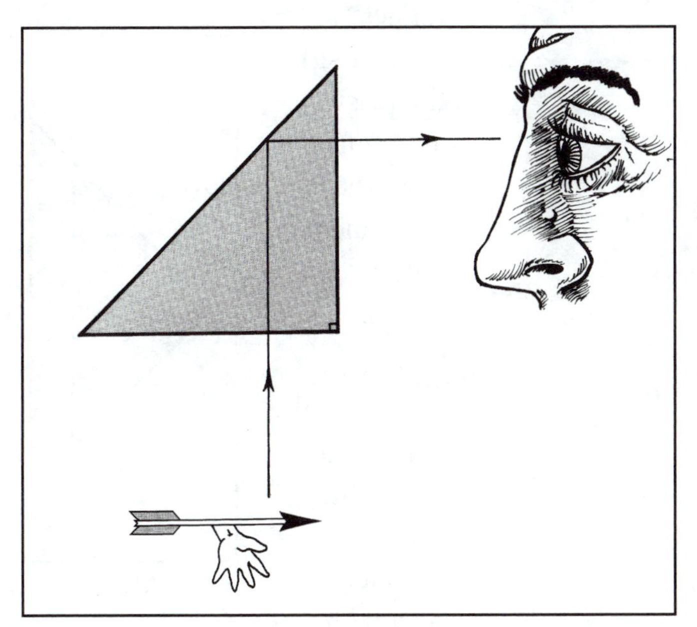

Figure 7.11

b. Two rays approach the prism from the bottom. Both are directed normal to the bottom surface. One ray enters the prism close to the left of the surface, the other enters close to the right of the same surface. Both rays emerge from the far side of the prism after a single internal reflection from the slanted surface. Sketch the paths of the two rays into and out of the prism.

c. The observer on the right side of the prism views the object below the prism through the prism. Where and with what orientation will the image be seen?

d. The prism is tilted by a small angle β as the object is viewed through the prism. How does a tilt affect the image location?

e. A total internal reflection at the slanted surface is required for the uses of this type of prism. What range of angles for the incident ray, as measured with respect to the normal to the bottom surface of the prism, will undergo total internal reflection at the slanted surface.

7.4.5 Pairs of 45°, 90°, 45° prisms are used in binoculars. In that application one prism is in the orientation shown in Figure 7.12. The other prism faces the opposite direction and is located lower by half the height of the prism. For the purposes of this problem you are asked to analyze the refractive properties of only one prism. Figure 7.12 shows the orientation of an object, the prism, and an observer. One of the rays from the object to the observer is also shown. The index of refraction of the glass is 1.52.

a. For the incoming ray shown, which need not be parallel to the right face of the prism, calculate the path of the ray as it enters the prism, is reflected twice inside the prism, and emerges from the same face it entered the prism.

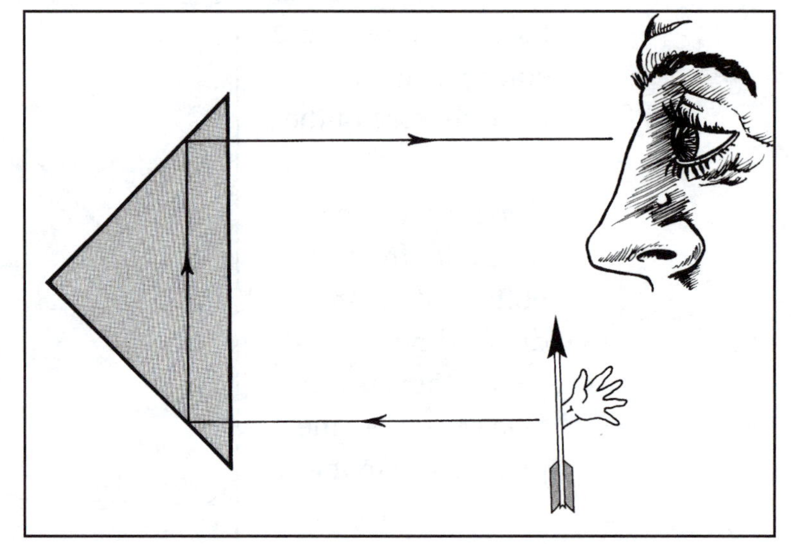

Figure 7.12

b. Two rays approach the prism from the right. Both are directed normal to the right surface. One ray enters close to the top of the prism. The other ray enters just above the center of the surface. Both rays emerge from the same face of the prism after two internal reflections. Sketch the paths of the two rays into and out of the prism.

c. The observer is on the right side of the prism and views the object located below the eye through the prism. Where and with what orientation will the image be seen?

d. The prism is tilted by a small angle β as the object is viewed through the prism. How does tilt affect the image location?

e. Two total internal reflections at the slanted surface are required for the uses of this type of prism. What range of angles for the incident ray, as measured with respect to the normal to the right surface of the prism, will undergo total internal reflection at the next two surfaces?

7.4.6 Figure 7.13 shows one of the possible orientations of object, prism, and observer. One of the rays from the object to the observer is also shown. The prism has the shape of an isosceles right triangle. The index of refraction of the glass is 1.52.

a. For the incoming ray shown, which makes an angle θ with the normal to the bottom face **A** of the prism, calculate the path of the ray as it enters the prism, is first reflected at side **B**, is reflected once more at the long face, and subsequently emerges from side **B** of the prism.

b. Two rays approach the prism from the bottom. Both are directed at the same angle θ with respect to the normal. One ray enters the prism closer to the right edge of the face **A** than the other. Both rays emerge from side **B** of the prism after two internal reflections. Sketch the paths of the two rays into and out of the prism.

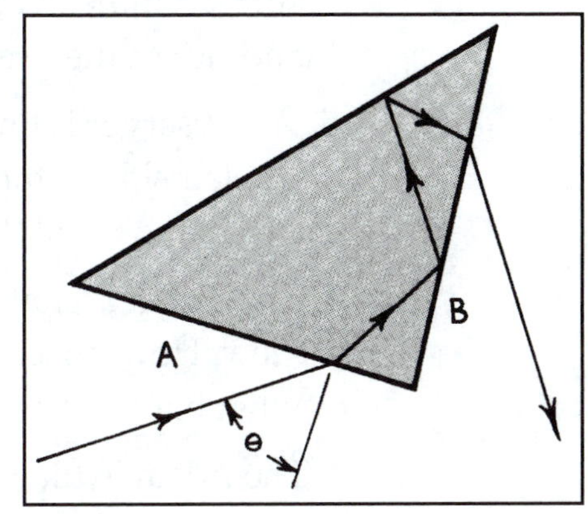

Figure 7.13

c. An observer is on the right side of the prism and views the object below the prism through the prism. Where and with what orientation will the image be seen?

d. The prism is tilted by a small angle β as the object is viewed through the prism. How does a tilt affect the image location?

e. A total internal reflection at the slanted surface is required for the uses of this type of prism. What range of angles for the incident ray, as measured with respect to the normal to the bottom of the prism, will undergo total internal reflection at the slanted surface?

7.4.7 A 45°, 90°, 45° prism is used in the orientation shown in Figure 7.14. To save on the cost of materials, on weight, and on the limitations imposed by the refractive index, the bottom portion of the prism is absent. The remaining optical device is known as a dove prism. Take the index of refraction of the glass as 1.52.

a. For the incoming ray shown, which need not be parallel to the top face of the prism, calculate the path of the ray as it enters the prism, is reflected at the upper face, and emerges again on the far side of the prism.

b. Two rays approach the prism from the left. Both are directed parallel to the top surface. One ray enters the prism close to the top surface, the other enters close to the bottom surface. Both rays emerge from the far side of the prism after a single internal reflection from the top surface. Sketch the paths of the two rays into and out of the prism.

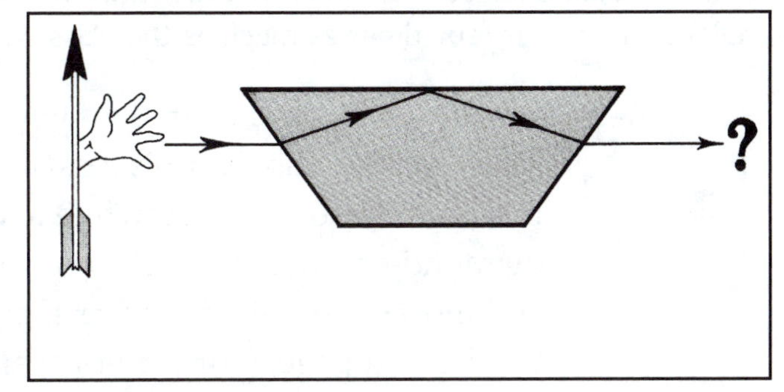

Figure 7.14

c. Prisms are three-dimensional objects. Two distinct rays can be parallel and enter the prism side by side at the same distance from the top surface and exit at the far side. Sketch the paths of these two rays into and out of the prism. Do they cross each other's paths?

d. An observer is on the right side of the prism and views the object on the left side of the prism through the prism. Where and with what orientation will the image be seen?

e. The prism is slowly rotated along an axis parallel to the direction of view. How does this rotation affect the image orientation? Specifically, what happens at 45°, 90°, and 180° rotations?

f. The prism is tilted by a small angle b as the object is viewed through the prism. How does a tilt affect the image location.

g. A total internal reflection at the top surface is required for the uses of this type of prism. What range of angles for the incident ray, as measured with respect to the top surface of the prism, will undergo total internal reflection at the top surface?

Optical materials used for lenses and prisms are characterized by an index of refraction. Glass types used for optical purposes are available with indices of refraction as low as 1.48 and as high as 1.90. The optometrist may choose to prescribe a set of glasses of a specified focal length by using lenses with small radii of curvature and low index of refraction, or by using lenses with more gentle (larger) radii of curvature and a high index of refraction. For very nearsighted or very farsighted individuals, the solution is to use both high index of refraction and small radii of curvature. Two facts work against using high index of refraction glass under normal situations. First, a higher index of refraction goes hand in hand with a larger variation of the index of refraction with wavelength. Second, a high index of refraction material reflects more light from its surfaces. The surfaces of glasses with index of 1.8 reflect twice as much light as the surface of glasses of index 1.5. You can easily see that very near-sighted people seem to have brighter looking glasses than people that are less visually challenged.

NOTE: Diamonds have an index of refraction of 2.42 and therefore have an even higher reflection—about four times as much as the glass with index 1.5; hence the brilliant sparkle.

7.4.8 If you were a jewel thief, then you would know that the longer it takes for a victim to discover his loss, the easier it is to get away with a theft. It is therefore often worthwhile to make an accurate copy of a large diamond-studded bracelet to substitute for the stolen bracelet. Glass and most other natural and artificial gems are cheaper than diamonds. On the next page, four materials are listed as choices for sparkling displays of wealth.

Index of refraction for	blue	yellow	red	hardness
sapphire (aluminum oxide)	1.776	1.769	1.765	excellent
lead glass (mostly PbO)	1.945	1.923	1.901	poor
diamond (carbon)	2.437	2.419	2.409	best
rutile (titanium oxide)	2.695	2.612	2.572	medium

a. Explain which of the materials listed in the table on the previous page will have the strongest reflection of light.

b. Which one spreads the spectrum most, that is, which is the most colorful?

c. Which of the materials listed would make the best fake diamond?

d. Which of the materials listed should you choose for a gift? Why?

Actually it is not only thieves who are clever. The rich who wear the diamonds often also own and wear pastes (copies), except in places where security is extremely tight.

7.5 MIRRORS AND LENSES

The above problems on reflection and refraction of waves are restricted to flat surfaces. However, a large fraction of the population wears glasses to correct vision problems, and even mirrors are often made with curved surfaces. This section begins with problems dealing with curved mirrors, their uses, advantages, and disadvantages. The section ends with problems concerning lenses.

In the mathematical derivations of the expressions for focal length f of spherical mirrors and spherical lenses, one difference stands out. The calculation for the mirror characteristics requires only knowledge of the radius of curvature R of the mirror,

$$\frac{1}{f} = \frac{2}{R}$$

whereas the calculation of lens characteristics also requires a knowledge of the index of refraction n of the lens material,

$$\frac{1}{f} = (n-1)\left(\frac{1}{R_1} - \frac{1}{R_2}\right)$$

where R_1 and R_2 are the radii of curvature of the two lens surfaces. An important practical conclusion can be drawn from this difference. As long as the reflecting surface of the mirror is sufficiently smooth and highly reflective, the properties of the mirror are independent of the wavelength of the incident radiation. Transparent materials, like glass or plastics, bend the direction of a beam of light in accordance with Snell's Law. As a result all transparent materials bend each wavelength differently because the index of refraction changes with wavelength.

Mirrors are therefore used specifically where the wavelength range can be large. A drawback is that mirrors can be awkward to use. In many cases the observer blocks part of the light needed to see the object.

7.5.1 Older people are amazed at the small details young children can see with the unaided eye. A young child's eye has a more flexible lens that can reduce its focal length such that objects as close as 10 cm can be seen clearly. Most older people cannot see an object clearly unless it is a meter or more away. Once the print of newspapers becomes too small to read or the arms of the person are no longer long enough to hold a book far enough from the eye, the aging person may need reading glasses to place the image of a nearby object at a distance where the eye can focus on it. Extremely nearsighted people can see objects as close as a child can, but they are unable to see clearly anything farther away. They need glasses to form discernable images at close range from objects which are far away.

a. The criterion which determines how well one sees details is the angle the object subtends in front of the eye at that distance from the eye which the lens of the eye requires to focus a clear image on the retina. The diameter of a dime is 18 mm. If a child can clearly see the dime at a distance of 9.5 cm from the eye, then calculate the angle that coin subtends from the eye.

b. An older adult sees the dime clearly at 1.00 m or farther. What is the largest angle over which this adult will see the dime with the unaided eye?

c. For the adult to discern the detail that the child can see with the unaided eye, the adult must also hold the dime at 9.5 cm from the eye but must place a lens between object and eye to form an image of the object at 1.00 m from the eye. Calculate the required focal length of the lens.

d. By convention the magnifying power of a lens is calculated based on a near point of 25 cm, even though few older adults can see well at that distance. Calculate the conventionally accepted magnifying power of the lens of *part c*.

e. If the child of *part a* were to use the same lens as the adult, then how close to the eye could that child hold the dime and still see it clearly?

f. For this child what would be the magnifying power of the lens?

g. With and without the lens, who sees the most details on the dime? Assume that the child and the adult differ only in the range over which their eyes can focus.

7.5.2 Some bicycles have a rearview mirror mounted on the handle bars. Such mirrors are mounted about 60 cm from the normal eye position of the rider.

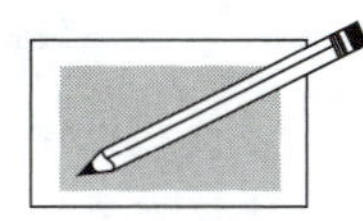

a. Recommend the type of mirror to be used—concave, convex, or flat. Explain what the rider would see in each case for a distant object, such as a car, and what the advantages and/or disadvantages are for each type of mirror.

b. One particular choice is a convex mirror with a focal length of -2.25 m. If a car is 100 m behind the the bicycle, then where will the image of the car appear as seen in the mirror?

c. The car is 1.90 m high. Calculate the angle subtended by the image of the car as seen in the mirror. Compare this angle to the angle subtended by the car as seen looking directly over the shoulder.

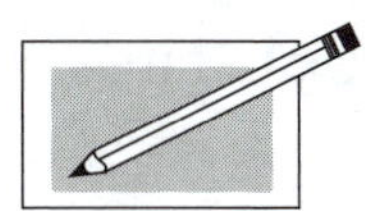

d. Explain why the car appears to be farther away when seen in the mirror compared to looking directly at the car, even though the image of the car in the mirror is much closer.

e. The car is approaching the bicycle from the rear and is now 50 m away. Calculate the change in the image position of the car as seen in the mirror. What has happened to the image size at the same time?

7.5.3 Spherical mirrors are used to look at our faces in detail. It may be to trim that whisker to perfection or to correct the curve of the eyebrow. The image in the spherical mirror is magnified and is located at a position where it is easy to see. (Imagine looking at your own nose through a magnifying glass without a mirror!)

Calculate the location and the radius of curvature of a mirror required to look at a small blemish on your eyelid. The conditions imposed by your eyesight are that you can see the blemish most clearly in a flat mirror held 27.4 cm from your eyelid, that the blemish appears five times larger than in a flat mirror, and that it is seen right side up.

Sherlock Holmes is usually depicted with his deerstalker hat on his head and a magnifying glass in his hand. Magnifying glasses are specified by their power. For example, 5 x or 10 x is supposed to indicate how much larger the user will be able to see the image as compared to the object. It is possible to achieve a large range of magnifications with any lens simply by changing object and image distance. Therefore it is important to understand why certain lenses are more useful than others as magnifiers.

7.5.4 The power of a magnifying glass is calculated on the assumption that the user clearly sees objects 25 cm from the eye. The role of the magnifying glass is then to form an image for the eye at that distance.

a. The object to be observed in detail is a splinter that has pierced the skin of Ophelia's finger. Its size is 3.4 mm by 0.052 mm. Calculate the angle subtended by the long side of the splinter if it is held 25 cm from the eye where, supposedly, it is clearly seen.

b. A +4.0 cm focal length lens is held close to the eye. Ophelia brings the finger with the splinter closer to her eye to get a better view through the lens. Calculate the distance from the splinter to the lens to obtain an image of the splinter at 25 cm from the lens. Calculate the angle subtended by the long side of the splinter as measured from the lens.

c. In calculations of the magnifying power of the lens it is convenient to ignore the fact that the eye and the lens are separated by a small distance, just as it is convenient to ignore that 25 cm is not an ideal viewing distance for everyone. Calculate the angular magnification achieved by the lens in terms of the ratio of the angle calculated in *part b*, as compared to the angle calculated in *part a*.

d. Ophelia may complain about the next step. We clamp her finger in a small but comfortable vise to hold it steady. The lens is then firmly mounted 4.4 cm away from the splinter. Calculate where the lens will project the image of the splinter. If a screen was placed at that distance and the light conditions were perfect, then the splinter could be seen on the screen. How long will the image of the splinter be? What angle will it subtend to an eye looking at the image from a distance of 25 cm?

e. Calculate the angular magnification in *part d* as compared to *part a*.

f. The lens in *part d* could have been mounted at 4.2 cm from the splinter in the finger. In that case what would be the angular magnification?

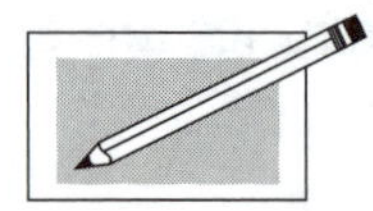

g. In *part d* is the statement "if… the light conditions were perfect." Explain what kind of light conditions would have to exist and how they can be achieved.

7.5.5 Antoni van Leeuwenhoek, a contemporary of Newton, was an expert at making extremely high-powered magnifying glasses and was the first to describe microscopic life. The best of his lenses, now preserved in a museum, has a magnifying power of 270 x. From his descriptions of small objects he saw, it is believed he achieved magnifications up to 500 x with his single lens glasses. Currently we rarely go beyond a magnification of 10 x with a single lens. Multiple lens systems called compound microscopes are used to achieve the high magnifications. Modern technology can reproduce and even improve on the lenses Leeuwenhoek made, but compound microscopes give better images and are easier to use. The earliest compound microscopes predate Leeuwenhoek by more than 50 years and were used by such famous earlier natural scientists as Galileo and Kepler. Their compound microscopes, however, did not come close to the capabilities of the Leeuwenhoek lenses.

a. What is the focal length of the 270 x magnifier made by Leeuwenhoek?

b. Assume that the lens is a symmetrical double convex thin lens made from glass with an index of refraction 1.60. What must be the radii of curvature of this lens?

c. What restriction does the radius of curvature of this lens place on the diameter of the lens? How does that in turn compare with the diameter of the pupil of the eye?

d. The object to be investigated by Leeuwenhoek was mounted on a sharp pin which could be turned and moved slightly. How far from the lens did the object have to be placed?

High quality camera lenses are assembled from a large number of individual thin lenses to correct for the widest variety of possible distortions. In spite of being complex, these lenses are specified by a single overall focal length and by a number which tells the user how much light will reach the film plane. The rim of the lens has information printed on it that includes the name of the manufacturer, the lens type, and then a string of numbers. One element of the string is the serial number that identifies the individual lens and is useful for insurance purposes. Following the serial number is the optical information in the form of an expression **1:a/b**, such as **1:5.6/135** or **1:1.9/40**. The number **b** is the focal length of the equivalent simple lens in millimeters. The number **a** gives the ratio of the focal length to the diameter of the equivalent simple lens. The number 5.6 therefore means that the focal length of the lens, 135 mm, is 5.6 times larger than the diameter. In the second example, the number 1.9 means that the focal length is only 1.9 times larger than the diameter of the lens. On a zoom lens the designation may be **1:3.5/28-70**, which means that any focal length between 28 mm and 70 mm can be chosen by the photographer. In addition, regardless of choice of focal length, the equivalent simple lens still has that same ratio of focal length to diameter of 3.5. The next few problems illustrate the meanings behind the numbers.

7.5.6 Three separate lenses for a 35 mm camera are labeled 1:3.5/28, 1:2.0/55, and 1:3.8/205.

a. For each of these lenses, state the focal length and calculate the diameter of the equivalent simple lens.

b. The width of the film used in the camera is 35 mm. The actual exposed area of the film is 24 mm x 36 mm. Each of the lenses, in turn, is used to photograph an automobile 4.5 m long. For each of the lenses, at what distance would the photographer have to stand from the subject in order to "fill" the long side of the film with an image of the automobile?

c. What angle of view does each of these lenses cover along the long side of the image?

d. The 55 mm focal length lens is considered the standard lens for a 35 mm camera. Why would the 28 mm and 205 mm lenses be considered "wide angle" and "telephoto" respectively?

7.5.7 Automatic features have made cameras easier to use by amateur photographers, but the serious photographer must be aware of the limitations of these tools. The questions in this problem explore some aspects of how much light gets onto the film in a camera and how this relates to object distance, magnification, lens diameter, and focal length. The questions are to be solved in strictly algebraic terms. The object has a height H and a width W, and is located at a distance d_o from the camera. The image is on the film, at a distance d_i behind the lens. The diameter of the lens is D and its focal length is f. Each mm^2 of the object has a brightness L, i.e. it uniformly emits that much light.

a. Derive an expression for the area on the film required to receive the light from the object.

b. Assume that the object emits light uniformly in all directions. What fraction of that light will enter the lens and form the image on the film?

c. Because the area of the image is smaller than the area of the object, the light will be concentrated by the lens. What fraction of the light from the object, expressed in terms of the variables d_o, d_i, and D, falls on the receiving area (the film)?

d. The image in a camera must always be a real image. What is the smallest possible value of d_i? Keep in mind that the focal lengths of camera lenses are usually less than 10 cm and that object distances are rarely less than 1 m. Expressed in percentage, how much will d_i vary under normal picture taking?

e. The photographer decides to include more background and therefore changes to a lens of half the previous focal length. The light conditions remain the same, the object distance remains the same, and the image must remain sharp on the film. To get the identical exposure (light per unit area) on the film, determine the effective lens diameter.

f. Camera stores sell close-up lenses that can be mounted in front of standard camera lenses. These allow the photographer to take pictures of objects that are only a few centimeters from the camera. As a result, a smaller portion of the object will be recorded on the film, but the magnification increases significantly. How will the extra magnification affect the illumination level at the image location? Express the answer in terms of the change in image size.

Getting a good photograph requires that a wide range of light intensities register on the film in the camera. Scenes for a photograph will have bright portions, dark portions, and a mix of colors. Proper exposure time and film type assure that enough light arrives from the dark portions of the scene to allow the chemicals in the film—silver bromide—to react. At the same time, the light from the bright portions of the scene must not be allowed to react with all the silver bromide in its path. This would cause the film to become uniformly dark on that portion of the image and no picture detail would be recorded.

Photographic film has a limited range of exposure that it can tolerate and modern cameras have been automated to seek that range. Bright scenes, such as skiers on a sunny slope, are handled by a small central portion of the lens and a fast shutter speed. The light intake is restricted to a time period of 1/250 s or less. When the light levels are less extreme, the camera computer or the knowledgeable photographer can decide to use a longer exposure time and/or a larger area of the lens (smaller f/number). When light levels are low the full area of the lens is used and the exposure time is stretched to the limit. With a hand-held camera an exposure time of longer than 1/25 s is likely to cause blurring because of a moving camera or moving subject.

When light levels are even lower two options exist. Flash photography is one option. A small, high-intensity light source at the camera momentarily illuminates the scene and the photograph is taken using the light from the flash lamp reflected back to the lens from the object. A major disadvantage of flash photography is that near objects will be much brighter than more distant objects—the $1/r^2$ law.

A second option is to use a high-speed photographic film. Photographic film is categorized by film speed, that is, the light requirement for a good exposure. Most films on the market now are rated ISO (International Standards Organization) 100. Other films commonly available are rated ISO 200, ISO 400, and ISO 50. Doubling the rating means that half the light is needed for a good picture. Higher film speeds are usually associated with higher cost and more grainy pictures. For professional photographers who can carefully control light levels and mount cameras on tripods for stability, lower speed films with a special fine grain are available.

7.5.8 The old-fashioned "point and shoot" camera is a simple device. The lens is fixed (no adjustment for distance possible). The lens is small in diameter compared to the focal length (approximately f/11). While it is inexpensive to make, it also permits a good range of object distances to be "in focus" on the film. The shutter speed is also fixed at approximately 1/25 of a second, short enough to allow for slight movement of camera or subject. With a film speed of ISO 50, the design combination satisfies the requirement of taking recognizable pictures of mom, dad, and the baby sitting on a bench in a sunny location in the garden. However, an error of a factor of two in light intensity results in a poor picture.

Better quality cameras have lenses which can be adjusted in two ways. The lens as a whole or some of its elements can be moved to modify the lens to film distance in response to object distances that are not just "far away" compared to the focal length. Another adjustment can be made to a diaphragm behind the lens. This adjustment allows the photographer to choose how much of the area of the lens is to be used. The technical term for the diaphragm settings is aperture.

a. The choice of diaphragm settings on a camera for serious photographers ranges from full use of the lens at possibly f/1.4 to smallest partial use at f/22. Standard steps marked on the lens are f/16, f/11, f/8, f/5.6, f/4, f/2.8, and f/2. Calculate the increase of light allowed through the lens between each of the steps listed.

b. Shutter speeds are also manually adjustable in higher quality cameras. The sequence is typically: 1/2, 1/5, 1/10, 1/25, 1/50, 1/100, 1/250, and 1/500 seconds. If the light conditions for Petunia playing with her dolls were correct at f/16 and at 1/25 s, then calculate the recommended shutter speed if the the diaphragm of the lens had been set to f/4. Why is that combination more appropriate for taking a picture of Petunia at bat at a softball game?

c. It is now late afternoon instead of midday and the light levels have dropped by a factor of four. How should the diaphragm and/or shutter speed be adjusted to take a picture of Petunia playing with her dolls?

7.5.9 Improvements in manufacturing techniques and the low cost of simple computer chips have made automatic features available in even low-priced cameras. The lenses are still small but a small light sensor measures the available light and adjusts the shutter speed to match the exposure time required. If the computer chip decides that the light levels are too low to keep the shutter speed at 1/25 s or less, a small light flashes to alert you to use the flash.

Flashlamps are annoying to the subject and sometimes of absolutely no use. A bit of realistic thinking about photographing musicians on stage at a rock concert is in *part c*. Flashlamps are rated according to their total light output, combining the brightness and the duration (1 ms or less) of the flash. A simple camera with a built-in flash and ISO 100 film is an adequate match to photograph a scene at a maximum distance of 5 m. Under the same conditions the professional flashlamp can provide enough light to illuminate a subject up to to 10 m away.

a. The illumination reaching an object from a small source like a flash obeys the ubiquitous $1/r^2$ law. The light from the object illuminated by the flash that returns to the camera lens also obeys the $1/r^2$ law. Find the ratio of the light outputs of the flashlamps of the professional vs the simple camera described above.

b. The exposure time required for the family photograph in the garden on a sunny day is 1/25 s. The same camera is used indoors with a flashlamp. The duration of the flash is 1/1000 s. How bright is the light from the flashlamp at the subject compared to the sun?

c. The simple camera referred to in the introduction to these questions can be reasonably assumed to have lens designated as f/4. The highest film speed available to the casual photographer is rated as ISO 400. The photographer might even be able to borrow a camera with a lens rated at f/2.0. Calculate the distance the subject can be from the flash and the camera to get an adequate exposure. Calculate the distance from the rock concert stage that the photographer would have to sit in order to get a reasonable photo.

NOTE: There is a major difference between having the flashlamp at the camera and having the flashlamp near the subject and the camera itself far away. Think about it.

There are hundreds of types of glass available from manufacturers. Some are more scratch resistant than others, some have higher and some have lower average indices of refraction. For some types the index of refraction changes more rapidly with wavelength than for other types. For the designer of optical systems each of these characteristics is important. Glass types are specified by a six-digit code. For example, the flint glass type SF-6 has the code number 805254. The crown glass type K-10 has the code number 501564. The first three digits of the code reveal the index of refraction at a characteristic wavelength in the yellow region of the spectrum. For the flint glass it is 1.805; for the crown glass the index of refraction is 1.501. The last three digits indicate how the index of refraction varies with color in the form $(n_{yellow}-1) / (n_{blue}-n_{red})$.

For the aforementioned flint glass the ratio is known as the Abbé factor. It is measured as 25.4. For the above crown glass the Abbé factor is 56.4. The wavelengths used for calibration of the glass in the blue, yellow, and red are respectively 486 nm, 588 nm, and 656 nm.

7.5.10 A lens designer gives the specifications for a plano-convex lens to be ground and polished for use in a camera. The focal length is to be 75.00 mm. There is a choice between two readily available types of glass from manufacturers. One is a flint glass designated as SF-11 and specified by the number 785258. The other is a crown glass, BK-7, specified by the number 516641.*

a. State the index of refraction of these two glasses in the yellow part of the spectrum.

b. Show that the indices of refraction in the blue and red for the flint glass at 1.80645 and 1.77599 respectively are consistent with the specification number quoted above.

c. The index of refraction for the crown glass is 1.52238 for blue light. Calculate its index of refraction for red light.

d. The lenses are ground for the average index of refraction (yellow). Calculate for each of the two glass types the radius of curvature required to achieve the focal length of 75.00 mm.

e. The lens is to be used in a camera to photograph a flower 82 cm from the lens. The flower has a diameter of 22.0 cm. How large will the image be in the yellow region of the spectrum for each glass type?

f. The sharpest images of the flower will be at different distances from the lens for different wavelengths. Calculate for these lenses how far from each other the blue and red images will be. Which glass type would you recommend as the most advisable? Why?

7.6 OPTICAL INSTRUMENTS, MULTIPLE LENS SYSTEMS

7.6.1 The obvious parts of a slide projector are (a) a light source to illuminate the slide to be shown, (b) the slide, and (c) the lens which projects the slide image as a much larger image on a screen. Two less obvious elements of a slide projector are the requirement for a lens to concentrate the light from the source on the slide and an electrical fan to keep the slide and other critical parts of the projector at an acceptable temperature. The following calculations will explore the system in more detail.

*Data from the Schott Glass Company.

a. A typical focal length for the projection lens system is 100 mm. The normal size of a picture in a transparency is 24 mm by 36 mm. For comfortable home viewing you might want the picture on a screen to be 120 cm by 80 cm. Calculate the slide-to-lens and lens-to-screen distances.

b. Assuming that all the light from the slide is actually projected onto the screen, calculate the relative brightness of the image on the screen as compared to the light levels on the slide.

c. The slides are to be shown in a partially darkened room with a background illumination level of 5 lm/m^2. For good viewing the image on the screen must be at least 20 times brighter than the background levels. Assuming an ideal efficiency of the lenses, calculate the minimum light level at the slide itself.

d. The light source used in good-quality slide projectors is the hot filament of a quartz halogen lamp. It is efficient in converting electric power to light, and it has a good output over all the colors of the spectrum. To keep the calculations simple, assume, as in Figure 7.15, the light source in the slide projector to be a point source radiating equally in all directions. Also assume the illumination at the site of the slide is required to be a circle 44 mm in diameter. If there were no mirrors or reflectors to direct the light toward the slide, then what would the distance between filament and slide have to be to keep the illumination at the slide uniform to within 5%? Under those circumstances what fraction of the light from the lamp will actually reach the slide?

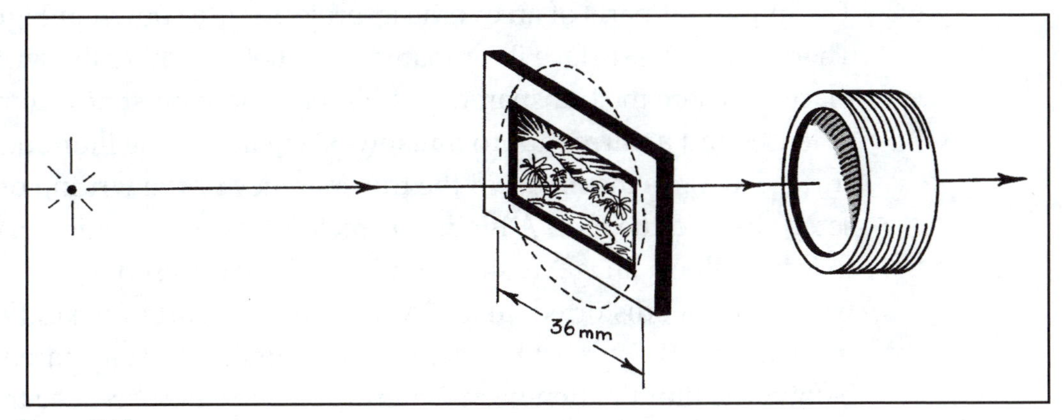

Figure 7.15

e. The projecting lens has an effective diameter of 29 mm and you have calculated in *part a* how far it must be from the slide. In this primitive arrangement what fraction of the light passing through the area of the slide will continue in the direction of the lens, pass through it, and ultimately reach the screen?

f. Two improvements are incorporated into slide projectors to increase the brightness of the image without having to use brighter lamps. Figure 7.16 shows that a spherical mirror is mounted behind the lamp to send a portion of the previously unused light back toward the lamp to help illuminate the slide. In addition, a convex lens is placed between lamp and slide. After allowing the light from the lamp to expand over an area greater than the slide, this added lens refocuses an image of the lamp at or close to the projecting lens. The slide is positioned between this refocusing lens and at the previously calculated distance (*part a*) from the projecting lens. What fraction of the output of the lamp will go to the slide this time? How much of that, in turn, will go to the projection lens?

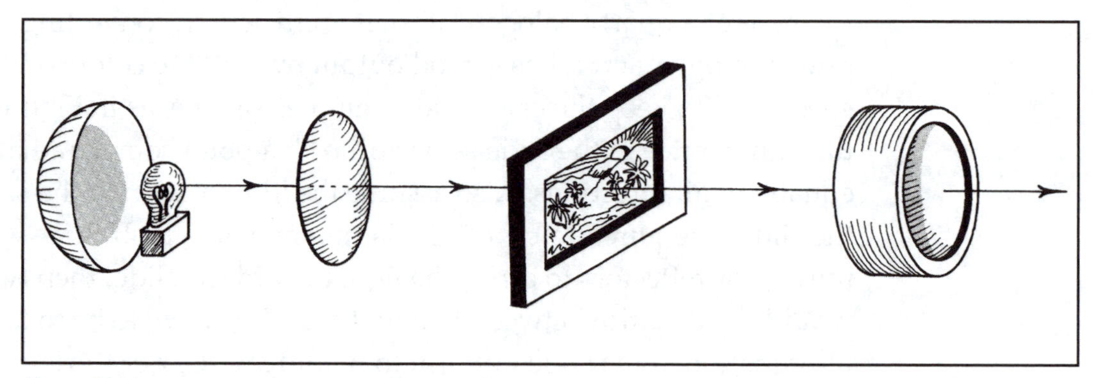

Figure 7.16

7.6.2 The important parts of an overhead projector are shown in Figure 7.17. There must be a surface illuminated from below where the writing takes place or where the transparency is placed. There must be a combination of lenses and a mirror approximately 30 cm above the illuminated surface to focus the image of the transparency or the writing onto a screen behind the user. A highly corrected lens combination is needed to form a sharp image. A simple spherical mirror used at an angle would give an distorted image. A flat mirror is part of the deflection mechanism. There must be a lamp to illuminate the transparency from below. The illumination needs careful consideration because the light eventually must be aimed at the focusing lens. This requirement is satisfied by using lenses and mirrors under the transparency to direct the light from the lamp to the focusing lens. The image of the lamp has to be focused to within a few centimeters of that focusing lens.

Try to take a close look at an actual overhead projector before starting the calculations below. Notice how the light from the illuminated surface proceeds from transparency to the focusing lens and reflector assembly and subsequently to the screen.

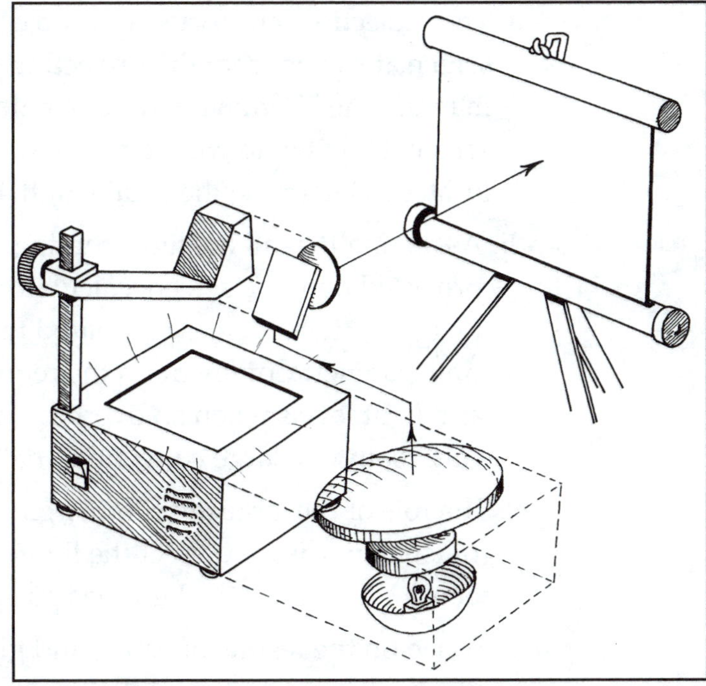

Figure 7.17

For the purposes of the numerical parts of this problem, the illuminated area on which the transparencies lie is 25 cm by 25 cm. The distance from transparency to mirror and projection lens is adjustable between

30 cm and 40 cm. The distance to the screen is typically 2 m or more. Inside the box which holds the light source and the plate on which the transparency rests there must be lenses and/or mirrors to distribute the light properly. The light must be spread evenly to illuminate the transparent plate and then it must be directed toward the mirror and the projecting lens. In some overhead projectors the arrangement is as shown in Figure 7.17. A spherical mirror below the lamp returns a part of the light to the filament. A converging lens of focal length 13.8 cm spreads the light from the filament to a large (approximately 35 cm diameter) converging lens, focal length 16 cm. This converging lens is located just below the transparent plate upon which the transparency rests. Also assume that the light source is a point source, i.e. negligibly small. This is the arrangement to be analyzed. In actual overhead projectors the large diameter positive focal length lens is a Fresnel lens to save space, weight, and cost. This is the reason for the coarse texture of the surface where the transparency rests. In some overhead projectors there are mirrors to fold the light path into a smaller space, again to make the machine lighter and more compact.

a. The projection lens focuses the object—the transparency—onto the screen at 2.25 m from the projecting lens with the requirement to magnify the 25 cm wide transparency to a 1.60 wide image on the screen. How far above the transparency must the lens and mirror be located? Determine the focal length of a projecting lens.

b. Assuming that all the light coming from the transparency is actually projected onto the screen, calculate the relative brightness of the image on the screen as compared to the light levels at the transparency. Comment on the requirements for the projection lamp and light levels when an overhead projector is used in a small classroom vs a large convention hall.

c. The role of the large area converging lens located just under the transparency is to redirect the light from the source through the focusing lens after the light has passed through the transparency.

Based on the results of *part a*, and given that the focal length of the large area lens is 16 cm, calculate the effective position of the light source such that its image will be placed at the location of the projecting lens.

d. The actual position of the light source is closer to the lens than the calculation of *part c* implies. The box under the transparency would otherwise be too large. There is, as stated in the introduction to this problem, an additional converging lens, f=13.8 cm, a short distance above the lamp to create an image of the lens at the position required in *part c*. The actual distance from lamp filament to transparency is 15 cm. Calculate the position of the lens from the transparency to satisfy the stated conditions.

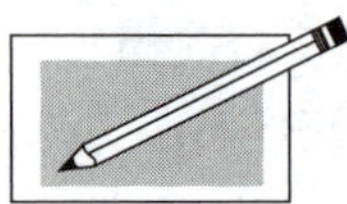

e. Why is it that the lamp in the projector cannot be focused on the transparency itself to maximize the brightness of the transparency? Would the transparency be evenly illuminated? What would be seen as an image on the screen?

7.6.3 Lens combinations can give counter-intuitive results. Two thin lenses of focal lengths f_1 and f_2 in contact with each other act as a single lens with a focal length f_c determined by

$$\frac{1}{f_1} + \frac{1}{f_2} = \frac{1}{f_c}.$$

For example, if f_1 were +20 cm and f_2 were -20 cm, then the equation predicts a combined focal length of ∞, in other words, the same as a flat piece of glass. However, as soon as the lenses are separated by a distance d, any positive focal length can be obtained for the combination depending on the choice of d. The combination of two or more thin lenses of arbitrary focal length can mathematically be considered equivalent to a single lens with a single well-defined focal length and an equally well-defined magnification. In practice multiple lens systems are used in cameras, microscopes, and telescopes because distortions introduced by one lens can be counteracted by the next lens so that improved images result. Unfortunately, the detailed calculations become so complex that computers must be used to optimize lens systems. A system of two thin lenses is still manageable and illustrates the principles.

The system to be looked at in detail consists of two lenses focal lengths +20.0 cm and -20.0 cm respectively separated by a distance of 4.0 cm.

a. An object 10 cm high is located 2.00 m to the left of the convex lens. Calculate the image position, the size, and the magnification of the object, ignoring the second lens.

b. The image formed by the convex lens becomes the [virtual] object for the second, concave, lens. Calculate the location of the image formed by the combination of the two lenses. Also calculate the size of the image produced and the total magnification of the system. Is this final image real or virtual?

c. For a single lens the absolute value of the magnification is the ratio of image distance divided by object distance. Show algebraically why the distance of the image from the second lens (final image distance) divided by the distance of the object from the first lens (initial object distance) will not give the absolute value of the magnification for a two-lens system.

d. Use the image position as calculated in *part b* and the object position as stated in *part a* to find the total separation between object and final image. Determine the focal length and location of a single lens which will give the same magnification as the two lens combination and which also leaves object and final image in the same place.

e. Sketch in a single diagram the position of the object, the original lenses, the single equivalent lens, and the final image.

7.6.4 Binoculars are really two telescopes mounted side by side, one for each eye. Each is adjustable to compensate for variations in eyesight and the distance between the eyes of the users. The cost of binoculars varies from $50 to $1000 or more. The quality of the image, which involves the elimination of distortions and color correction, plays the main role in cost. A reasonably good set of binoculars has three numbers in its optical specifications. The designation 8 x 50 means that the image is magnified by a factor of 8 or, from a different point of view, the object appears to be 1/8 of its actual distance from the observer. The number 50 in the expression 8 x 50 refers to the effective diameter of the objective lens. The third number is the angle of view in degrees which can be anywhere from 2° to 20°. The angle of view is simply the angle covered from left to right without moving the binoculars.

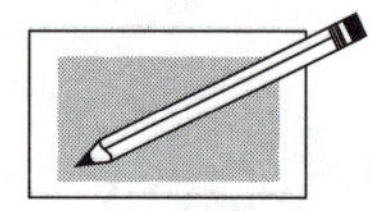

a. Binoculars are used to make distant objects look larger (or closer). At first glance more magnification should be more desirable. Explain why a magnification of 12 or more is impractical.

b. Why is it important to know the diameter of the objective lens? (Some binoculars are especially designed for night viewing. How can you tell?)

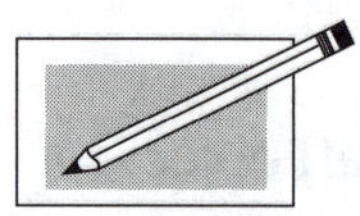

c. Tell why a large angle of view through binoculars would be preferred. Also explain why a narrow angle of view is almost always a consequence of high magnification.

d. In addition to angle of view, the specifications for binoculars also include a statement of the field of view. An example of such a statement might be that the field of view is 122 m at 1000 m distance. The numbers mean that as you look at a scene 1000 m away, you can only see a 122 m wide stretch. Show that this specification is the same as an angle of view of 7°.

e. Too large an angle of view will also cause trouble. Suppose your field of view through the binoculars were 500 m at 1000 m and the magnification is 8 times. This would mean that you would see a scene 500 m wide at a perceived distance of 125 m. What angle does this view represent? Does that make for good viewing?

8 PHYSICAL OPTICS

8.1 INTERFERENCE

Two optical phenomena are superficially similar in visual effect, though quite different in physical and mathematical complexity.

The simpler of the two, called Moiré patterns, may help you visualize the complex one, interference, that follows.

Moiré Patterns

One version of the simple effect is illustrated by the two sets of lines reproduced in Figure 8.1. To see the effect trace either one of the sets of lines on a piece of transparent paper or plastic. Place your tracing over the other set of lines and you will see a broad repetitive pattern of light and dark regions. The same effect can be achieved with other repetitive shapes, like circles, or it can be seen by looking through one row of equally spaced vertical fence boards at an identical row close by. There is also a home-based option. Look at a bright background through two long hair combs separated by a centimeter or so. You should see bright and dark patterns that shift as you move your head.

Figure 8.1

In the example of the black lines, the open space between the lines of the top pattern is aligned with the open space of the second pattern, or it can be aligned with the dark line below it. Because the width and separation of the printed lines are different for the two sets of lines, the superposition will form a new pattern.

In the examples of two rows of fence boards, or two combs, the separation between the boards or teeth may be identical, and the pattern is seen because of changes in the angle of view.

The pattern that is seen in both cases is called a Moiré pattern. There are many more examples in nature of these kinds and the same name, Moiré, is also used for a special weave in silk cloth.

8.1.1 Two long, identical combs are held in parallel, 0.85 cm apart with the teeth all pointing down. The left ends of the combs are aligned behind each other. The two combs are viewed through one eye that is 22 cm in front of the first comb, in line with one end of the combs.

The combs are standard fine hair combs. The teeth are 0.70 mm wide as seen from the side, and there is a 0.70 mm gap between the teeth.

When viewed as in Figure 8.2, the gap between the teeth allows some of the scene behind the combs to be seen. When the view is shifted, there will be directions in which the scene is blocked as the teeth of the second comb appear in the gap between the teeth of the first comb and then the view opens again as the teeth in the second comb are fully or partially obscured by the teeth of the of the first comb. The phenomenon repeats if the combs are long enough.

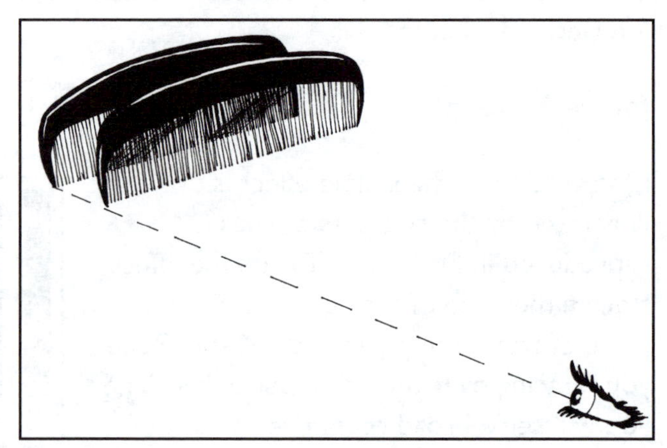

Figure 8.2

a. When viewed straight ahead, a tooth of the first comb completely obscures a tooth of the second comb. Determine the shortest distance, as measured along the first comb, at which a tooth of the second comb will be again hidden exactly behind a tooth of the first comb. Determine the angle, as seen by the observer, at which this occurs.

b. There will be repetitions between light and dark regions. Prove whether this repetition is at equal distance intervals or at equal angular intervals, as measured from the side.

8.1.2. In one pattern of black lines on a transparency, the width of the lines is 3.8 mm while the blank space between the lines is 3.3 mm. A separate transparency has a set of lines with a width of 4.5 mm separated by 3.2 mm. The two transparencies are laid on top of each other on a piece of white paper. Calculate the distance between adjacent centers of the darkest area formed by the superposition of the patterns of the two transparencies.

FYI: Patterns like this are mathematically equivalent to beats in wave propagation and sound. In this problem we are dealing with repetitions in space, while beats are repetitions in time.

The wave character of light is the origin of light and dark patterns that can remind you of Moiré patterns, but the scale of the effect in terms of the size of objects involved is orders of magnitude smaller. Differences in path of fractions of micrometers become crucial instead of centimeters or even meters. As before, the patterns change with the angle of viewing and repetitive, uniformly spaced reflecting surfaces are an important aspect to form the regular optical interference patterns. Irregular optical interference patterns are also easily seen. These are most often encountered as the speckle patterns produced when a laser beam is reflected from a roughened surface.

All standard textbooks have the usual problems on interference due to two slits, multiple slits that become diffraction gratings, and the thickness of oil films on water that cause colored reflections. The problems here explore other interference effects.

For critical optical applications, such as precision telescope mirrors and lenses in laser applications, the optical surfaces have to be perfect on a scale measured in fractions of a wavelength. Slight bumps and scratches scatter precious bits of light or might give rise to spurious images.

Interference effects are a good way to find imperfections. The problem below uses the most simple example of interference, that between two flat surfaces at a slight angle. The method is equally applicable to curved surfaces, but the mathematics becomes more cumbersome.

8.1.3 The task is to look for imperfections on an ostensibly perfectly polished flat piece of glass. The method requires a previously made standard called an optical flat, the test piece, a thin wedge to separate the plates at one end, a single wavelength source of light (usually a sodium or mercury lamp), and a magnifying glass to observe the interference patterns.The Figure 8.3 shows the arrangement with the dimensions as indicated. The wedge can be a fine hair but most often it is a microscope cover slide, typically 0.14 mm thick. The test piece and the optical flat are typically about 10 cm long. The two good surfaces face each other and are separated at one end by the wedge. The observable interference takes place between the light reflected from the top surface of the test piece and the internally reflected light from the bottom surface of the optical flat.

If the test piece and the optical flat above it are perfectly flat, interference fringes can be seen when viewed from above.

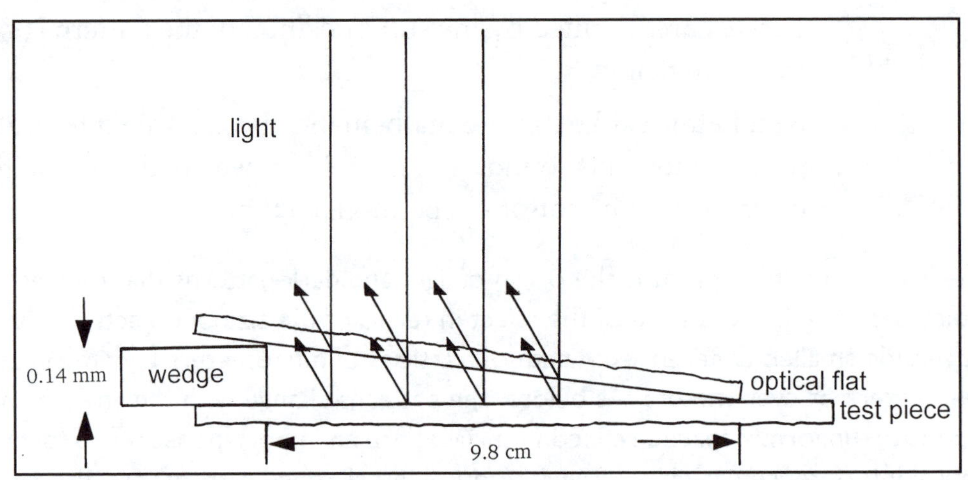

Figure 8.3

a. The arrangment is illuminated by a parallel beam of green mercury light at a wavelength of 546.1 nm. The light reflected from the upper surface of the test piece combines with the light internally reflected upward from the bottom of the optical flat to form interference fringes. Determine the horizontal separation between successive bright lines as viewed from above.

   Note that the two other surfaces of the pieces of glass play no role in the patterns formed. The pieces of glass are too thick and the untreated sides are rarely uniform enough to form interference patterns.

b. Figures 8.4a and 8.4b below, mirror images of each other and NOT drawn to scale, show patterns of interference lines that could be seen under magnification in a small segment under the arrangement described in *part a*. The two glass plates are in contact at the right, and the wedge is on the left side. Explain which, 8.4a or 8.4b, represents a dip and which represents a bump on the test piece. How deep or high is the dip or bump? Which would be easier to remove by additional polishing?

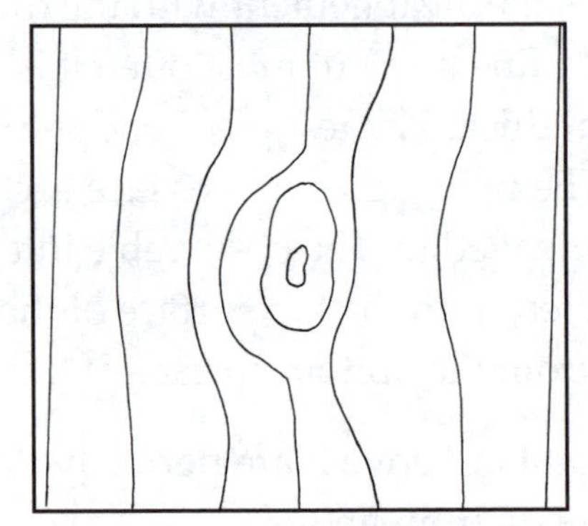
Figure 8.4a

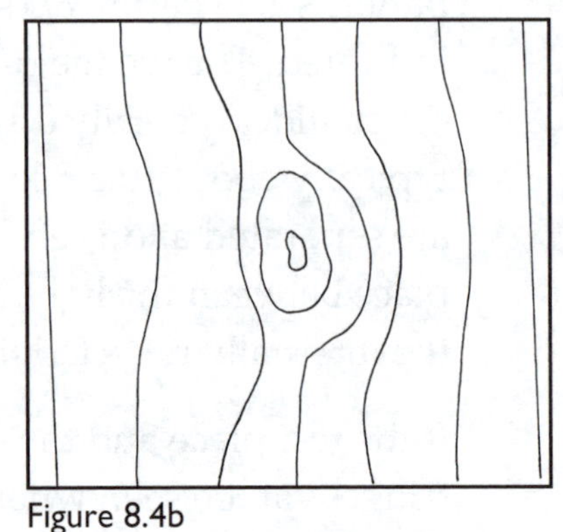
Figure 8.4b

8.1.4 Interference effects can be used to measure the index of refraction of gases or even the presence of gases. The schematic in Figure 8.5 shows a beam of monochromatic, parallel light coming from the left. A partially silvered mirror splits the beam into two beams at right angles to each other and, after two more reflections, what is left of the beams recombine. There is a sample chamber (cell) in the path of each beam. These cells can be evacuated or filled with a gas or a liquid.

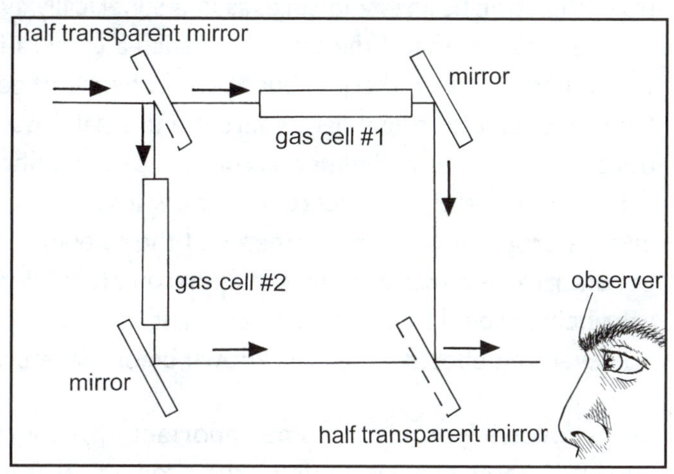

Figure 8.5

The system is so precisely manufactured and adjusted that, as long as the two chambers have the same fill, the beams recombine with zero path difference. In other words, there is not even a difference in phase between the beams.

Suppose now that the apparatus has been perfectly adjusted with cells #1 and #2 evacuated. Slowly, air or some other gas is allowed to enter cell #1. Because the gas reduces the speed of light beam in cell #1 ever so slightly as compared to cell #2, the two beams will be out of phase when they recombine. The magnitude of the retardation is a function of the length of the cell and type and density of the gas.

a. The length of cell #1 is 9.500 cm. The index of refraction of air at atmospheric pressure differs from exact unity by 2.765×10^{-4} at the 589.0 nm emission of a sodium lamp. Calculate the change in optical path length in the cell as air at atmospheric pressure is allowed to displace the vacuum.

b. How many wavelengths of light at 589.0 nm does the change in optical path represent?

c. The observer counts fringes as the air is slowly permitted to fill the cell. How many fringes, from light to light, can she count during the filling process? At what fraction of atmospheric pressure will the first complete change of one fringe have passed? In practice, this kind of measurement is done opto-electronically.

8.2 DIFFRACTION

The laws of geometrical optics work well as long as extreme details in image quality are not required. For example, the pinhole camera is based on the geometric principles that light travels in straight lines and spreads in a spherically symmetrical way from a source. The implications are that if the pinhole of such a camera is made small enough, then the image would become ever sharper but at the expense of getting an ever dimmer image. Experiments with bright light sources and small slits or pinholes prove otherwise. The image does indeed become dimmer as the size of the pinhole is decreased. However, as the diameter of the pinhole decreases below a specific threshold, the image size increases in an inverse proportion to the diameter of the opening. The threshold diameter at which the geometrical assumptions fail is at approximately 0.5 mm, about 1000 times the wavelength of light being used. Lasers are the best light sources to demonstrate these diffraction effects; however, the effects were well known before lasers were invented.

In the following problems some important physical constants are stated to the limit of their currently known values. In doing the problems it is important to round off these numbers to fit the other data in the problems.

8.2.1 The common helium-neon laser emits a bright red (632.816 nm) nearly parallel beam of light. The beam passes through a vertical slit with an adjustable opening and falls on a screen 4 m away as a spot 6 mm in diameter. The width of the slit is slowly decreased. As you observe the screen the spot suddenly shows fringes to the side, almost like a halo. As the slit opening gets smaller, the bright central spot grows dimmer while it stretches horizontally, but not vertically. The width of the central bright spot on the screen is defined as twice the distance from the center of the bright spot to the darkest region to either side.

a. At what slit width, in mm, will the spot on the screen spread from 6.0 mm to 6.3 mm?

b. Draw a graph based on calculations, plotting the width of the central spot vs. slit width. At what slit width does the spot have its narrowest dimension?

c. Sunlight too is almost parallel light by the time it reaches the Earth. Predict in detail what will be seen in a room if sunlight enters through a blackened window having an opening only 0.10 mm wide and the light shines on a white wall 3 m away.

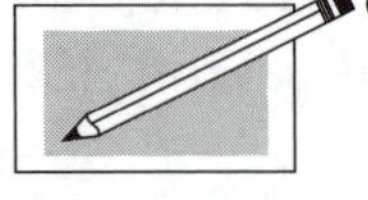

d. Hold your hand at arm's length away from your eyes. Look at a bright light through a gap between your fingers. Describe as carefully as you can the pattern of light you see in the gaps between your fingers as you change the width of the gap by squeezing your fingers together.

8.2.2 Satellite receiver dishes are common in the rural areas where it is expensive or impossible to be connected to cable TV systems. Currently there are two frequency bands at which satellites broadcast. The older ones broadcast at a frequency of 4 GHz, while more recently launched satellites use frequencies close to 12 GHz. All these satellites are in synchronous orbit above the equator, which means they orbit around the Earth at the same rate as the Earth rotates around its own axis. This in turn means that a rigidly mounted receiver dish pointing at a certain satellite remains pointed at that satellite at all times.

Each of these satellites must maintain a distance of 42.3 million meters above the center of the Earth. The radius of the Earth is 6.378 million meters. To avoid gravitational disturbances between satellites, they are a minimum of 1.48 million meters apart.

a. For someone living at the equator, such as in the northern regions of South America, calculate the distance from the dish to the orbit where a satellite is to located. Using the minimum separation between the satellites, use the Rayleigh criterion to determine the minimum diameter of the receiver dish that will pick up one of the satellites broadcasting at 4 GHz without interference from a neighboring satellite broadcasting at the same frequency. Repeat the calculation for two neighboring satellites broadcasting at 12 GHz.

b. Edmonton, Alberta, Canada is located 53.5 degrees north of the equator. Calculate the minimum diameter required there to avoid tuning in to two neighboring satellites at 4 GHz and two neighboring satellites at 12 GHz. At what angle above the horizon should the receiver dishes be pointing?

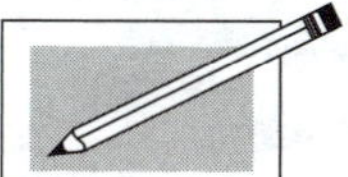

c. Explain why, under all the above conditions, satellite dishes should be well above the minimum required diameter.

d. It is to be expected occasionally that the sun will be on the same straight line with the satellite and the receiver. Reception deteriorates during that time because the sun emits radio frequencies just as abundantly as it emits light. Still based on the Rayleigh criterion, how long should the reception be seriously affected by the sun's interference?

8.2.3 Several unmanned satellites orbit the Earth for the purpose of studying crops, weather, and geographical details. Others are spy satellites. Objects the size of automobiles are clearly recognizable from these satellites whose orbits are 200 km to 300 km above the surface of the Earth. To distinguish one object from another details of the object must be visible. As a rough estimate the resolution should be ten times better than the size of the object.

a. What diameter lens must a (video) camera have in order to resolve two points 20 cm apart from an altitude of 200 km? Base your calculations on the use of a wavelength of 580 nm.

b. The same satellites also probe the Earth's surface at infrared wavelengths that pick out hot objects in cooler surroundings, such as forest fires, power plants, or the exhaust emissions of trucks. A wavelength range used by the Landsat 4 and Landsat 5 satellites for measurements in the infrared is 10.4 μm to 12.5 μm. What diameter of lens or mirror would be required if it were necessary to see the same detail in this infrared region of the spectrum as the (video) camera in *part a* can see in the visible part of the spectrum?

NOTE: There is a second difficulty with achieving high resolution in the infrared part of the spectrum. The intensity of infrared radiation from a hot object is usually much less than the reflected sunlight in the visible part of the spectrum.

8.2.4 Radar uses electromagnetic radiation in the centimeter wavelength range. It is a well established tool and is used by police to catch speeders and by pilots and navigators in aircraft and aboard ships to "see" ahead. There is a well documented example of biological "radar." Bats use sound waves instead of electromagnetic waves to "see" their prey. They emit bursts of ultrasound in the frequency range of 20 kHz to 40 kHz, which are out of range of human hearing. The waves are reflected by the insects they prey on. The reflections are detected and interpreted by the bats and they are able to find their prey. The speed of sound in air is approximately 330 m/s.

a. What is the wavelength at which bat "radar" operates?

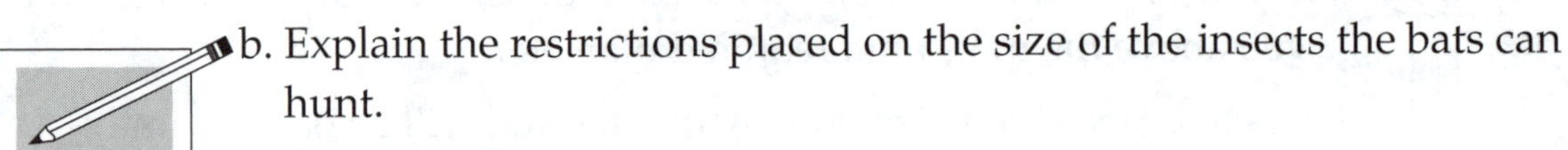

b. Explain the restrictions placed on the size of the insects the bats can hunt.

8.3 THE SPEED OF LIGHT

The speed of light can be measured to extremely high accuracy and is currently <u>defined</u> as 299 792 458 meters per second, with the second defined on the basis of the frequency of a specific atomic vibration (atomic clock). The meter has become a distance calculated from the definitions of speed of light and the second. It seems like a strange way to go about things, but those who must make extremely accurate measurements have found this to be the most reliable way to calibrate their instruments.

8.3.1 It is said that Galileo tried to measure the speed of light by standing on one hill holding a light source and having a servant stand on another nearby hill holding another light source. The light sources were likely candles in a box that could be covered. The idea behind the experiment was that at a certain moment Galileo would uncover his candle; the servant would see the light and immediately uncover his candle. Galileo would then note how long it took for him to see the flash of light in response to his own flash.

a. Suppose that the two hilltops in Galileo's experiment were 5.6 km apart. Using the currently accepted value of the speed of light, calculate the time interval Galileo would have needed to be able to measure in order to determine the speed of light.

b. The reaction time of the human body, even to start and stop a stopwatch, is approximately 0.1 seconds. It may be assumed that this was the time it took the servant to uncover his candle when he saw Galileo's candle. How does this reaction time compare with the time interval Galileo needed to measure?

8.3.2 The Danish astronomer Olaus Römer, working in the Paris Observatory in 1676, was able to predict the observed variations in the timing of the eclipses of one of the moons of Jupiter. He based his predictions on the contemporary knowledge of the distances from the Earth to the sun and from Jupiter to the sun, with the additional assumption that the speed of light had a definite value. In fact Römer used these astronomical observations to calculate the velocity of light. It was met with skepticism. Some aspects of his calculation are simplified in this problem but the principle is the same.

For the purposes of this problem, Earth revolves around the sun in one year in a circular orbit of radius 149.6 million km. Jupiter revolves around the sun in a circular orbit that has a radius of 778.33 million km. It takes Jupiter 11.862 years to make a complete circuit. Figure 8.6 shows Earth and Jupiter at closest approach and then where they are three months later. Earth will have moved along 1/4 of its orbit, while Jupiter has moved only a few degrees and could, for this calculation, still be considered to be in the same position.

a. Calculate the distance between Earth and Jupiter at closest approach. Also calculate their separation three months later.

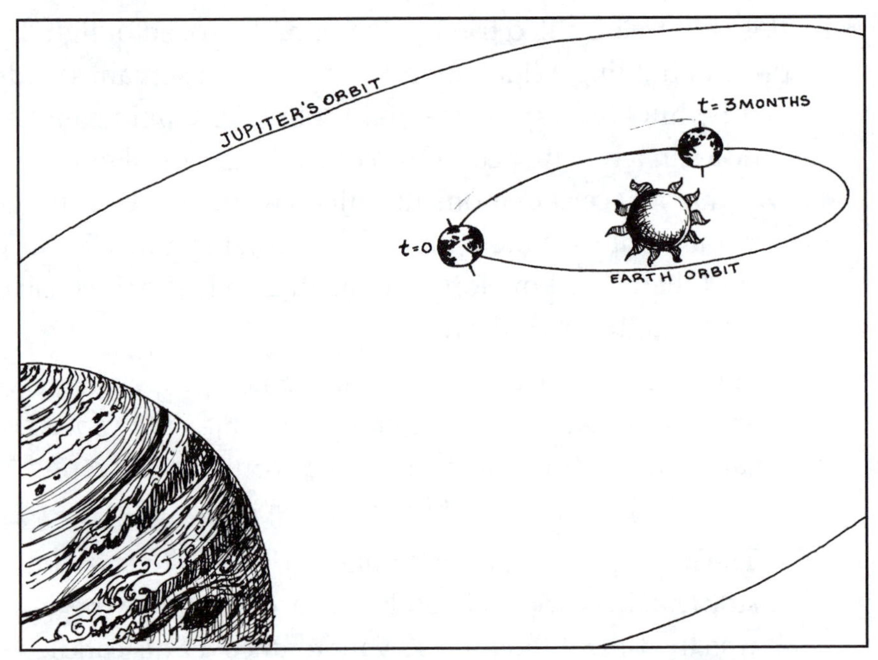

Figure 8.6

b. How much time would elapse in each of these cases before an event that happens on Jupiter is seen on Earth?

c. There are accurate clocks that move with Jupiter and can be observed from Earth. These clocks are Jupiter's moons that were first seen by Galileo. However, imagine you can read an actual clock on Jupiter. Suppose further that you can synchronize a clock on Earth with the clock you see on Jupiter at the moment of Jupiter's closest approach to Earth. After three months have passed, how much time will the observed clock on Jupiter have appeared to have gained or lost compared to the clock on Earth ?

d. **CHALLENGE:** Jupiter does move a significant distance in 3 months, a fact that was neglected in the calculations of *parts a, b,* and *c.* Include this fact now and recalculate more precise answers for *parts a, b,* and *c.*

8.3.3 Much international communication is now transmitted by beaming signals to satellites and from there back to other locations on Earth. Most communications satellites, including Canada's Anik satellites, are in so-called stationary orbits above the Earth's equator. They appear to be stationary because their orbital speed matches the revolution of the Earth. The laws of mechanics require that the orbits have a radius of 42.3×10^3 km measured from the center of the Earth. The radius of the Earth is 6378 km at the equator.

a. How long does it take for a signal to travel from the Earth's equator to the satellite and back? Would this be noticeable in a telephone conversation?

b. Compare the satellite transmission time with the time it takes for a radio (telephone) signal to travel halfway across the Earth but along the Earth's surface. Is that delay noticeable?

8.3.4 For many practical applications it is enough to know that the speed of light is about one foot (30 cm) per nanosecond. In space exploration, in particular for the corrective signals that must be sent to space probes as they approach a distant planet or when they are brought to a rendezvous with a comet, knowledge of distance and signal delay becomes critical, as a simple calculation will show.

Saturn is in an orbit approximately 1430 million km from the sun. Earth is only 150 million km from the sun.

a. At closest approach between Saturn and Earth, approximately how long does it take for a signal from Saturn to reach Earth?

b. Currently the most precise way of determining the distance to the moon is to send a light pulse to the moon and then to measure the time it takes for the reflected light to return to Earth. A space probe, such as Voyager, has allowed similar measurements to be made with radio signals. The speed of light and radio signals is exactly 299 792 458 m/s. Had it been erroneously believed to be 299 792 456 m/s, by how many km would the calculation of the distance from Earth to Saturn be in error?

8.3.5 To make room for the "information highway" more and more communications traffic is being channeled through optical fibers instead of electrical cables and microwave relay towers. These optical fibers are pulled from ultra high purity specialty glasses. Fibers with an index of refraction of 1.452 are used for a certain wavelength range.

a. How far ahead would a light pulse traveling in air be as compared to a pulse starting next to it but traveling in an optical fiber over a distance of 2.4 km?

b. Timing is extremely important for optical communications. Information is carried as a series of light flashes, with repetition rates on the order of 10 000 million per second. The duration of each flash is the same as the duration of the "off" period between flashes. How long does it take for such a flash to pass a certain point in air? In the fiber?

c. Calculate the distance a pulse must travel in air to get ahead by 100 of these pulses compared to a signal traveling through the same length of an optical fiber. In the applications of fiber optics the difference in travel time inside and outside a fiber can be ignored because the light outside dissipates much quicker than the guided light in the fiber. There are, however, serious technical problems strictly inside the fiber. Different wavelengths travel at different speeds, causing a spreading of the pulse. In addition, even inside a thin fiber, some fraction of the light bounces back and forth along the walls of the fiber and is also retarded compared to a "straight through" beam.

8.3.6 Modern surveying instruments determine distances by measuring the precise length of time it takes for a flash of light from the surveyor's instrument to travel from that instrument to the reflector held by an assistant and back again. The light source is a laser or light emitting diode of wavelength 820 nm. Corrections are made by internal computers for variations in the index of refraction of the air because atmospheric conditions will noticeably influence the speed of light.

a. How precise must the timing mechanism (internal clock) be for the distance to be correct within an accuracy of 1 cm?

b. Approximately how many wavelengths of light does that accuracy represent? The reflector on the rod carried by the assistant is a corner cube prism of the type described in Problem 7.3.6. Why is this type of reflector so useful in this application?

8.4 POLARIZATION

There have been problems about waves using the concepts of frequency, velocity, and wavelength. Waves are most easily visualized as water waves rippling along the surface of a pond. However, waves can be transmitted in or along materials in different ways. A stretched string is a good example. The string can be plucked sideways to generate a wave that travels along with alternately left and right deviations from the rest position. The string can also be plucked vertically to generate up and down deviations from the rest position. Less obvious, but equally feasible, is to jerk the string along its length. The result will be a wave in which the constituents of the string move back and forth along the string. A fourth type of wave is more difficult to see on a string but can easily be seen on a ribbon or a rope ladder. The wave is started by a twist and the twisting motion travels along the string or ribbon. The first two types of waves are called transverse waves and are characterized by displacements of the medium (e.g. the string) in a direction perpendicular to the direction the disturbance is transmitted. The next type is called a longitudinal wave. It is characterized by displacements of the medium back and forth along the direction the disturbance is transmitted. The last example is a torsional wave.

The above examples are but some of the variety of waves that can be generated and transmitted. Also note that some media restrict the types of waves they can transmit. For example, sound can only be transmitted as longitudinal waves in gases, while light and radio waves can only be transmitted as transverse waves. Each mode of vibration can also have its distinct speed of transmission. Problem 5.3.1 made use of the difference in wave velocities between shear (transverse) waves and compressional (longitudinal) waves.

Just as there are means of generating each type of wave independently, it is possible to make detectors that react to only one mode of wave transmission, or to make filters that allow only one mode to pass through a material. Polaroid filters are an example.

8.4.1 One property of transverse waves has biological applications. A beam of polarized light is launched into a glucose solution with a vertical direction of vibration. It is then found experimentally that the direction of vibration rotates as it progresses through the solution, as if spiralling. How much it turns depends on the concentration of glucose. Furthermore, one form of glucose with the simplified formula $C_6H_{12}O_6$ is called d-glucose or dextrose or corn sugar. It rotates the plane of polarization in a direction similar to the thread of a conventional right-handed screw. L-glucose, also called levulose or fructose, is a second form of $C_6H_{12}O_6$. It rotates the plane of polarization like a screw that needs to be turned in the opposite direction of a conventional screw.

Fructose is a natural form found in fruits and honey. Dextrose is formed together with levulose in the chemical breakdown of starches. Glucose is but one of many organic compounds that nature creates with a preferred direction of rotation, but that chemistry creates with a random direction of rotation. There are significant nutritional differences between left-handed and right-handed forms. In the laboratory they are distinguished using polarized light.

The total amount of glucose can be determined by methods of quantitative chemistry. Where the purity and origin of food additives is important, the presence of d-glucose in sweeteners is an indicator that artificial rather than natural ingredients have been used. The direction and total angle of rotation of polarized light is the standard way of determining the artificial components. Handbooks for people working on food additives list the "specific rotation" of sugars and amino acids.

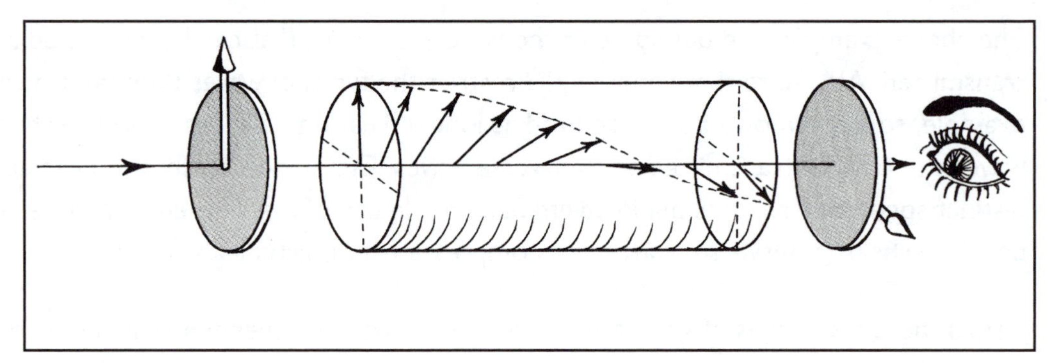

Figure 8.7

A device called a polarimeter (Figure 8.7) is used to measure the angle through which the polarization of the light has turned. The measured angle of rotation θ will depend on the concentration of the sugar d stated in g/cm^3, the length of path l in cm the light has to take through the solution, and the specific rotation ρ. The relation is

$$\rho = 10\theta \;/\; ld$$

When we deal with a mixture, each component will have its own concentration sand make a contribution to the rotation. A left-handed rotation of one component will counter a right-handed rotation of another component.

a. A sweet-tasting liquid has been analyzed to contain pure glucose with a concentration of 0.12 grams per cubic cm. As dextrose the plane of polarization turns clockwise with a value of the specific rotation of +52.5°. If on the other hand the glucose were of the levulose type, then the specific rotation would be –51.4°. Calculate the angle and direction of rotation expected in either case if the the polarized light passed through a 10 cm length of such a solution.

b. Calculate the angle and direction of rotation of the polarized light if the solution had been half dextrose and half levulose.

c. Calculate the angle and direction of rotation of the polarized light if the solution had been 30% levulose and 70% dextrose.

d. The total concentration of the two components is still 0.12 grams per cubic cm. The polarized light is passed through 20 cm of solution. The plane of polarization is found to have been rotated by –3.7°. Calculate the concentrations of levulose and dextrose.

8.5 FREQUENCY IN ELECTROMAGNETIC RADIATION

We used the speed of light, polarization, and the wavelength of light to solve some problems. A few calculations are useful to see where optical frequencies are located in the electromagnetic spectrum as compared to sound, microwaves and radio frequencies.

8.5.1 Ultraviolet, visible, infrared, and radio waves all travel with the speed of light, but their wavelengths and frequencies differ greatly.

a. Calculate the frequency of ultraviolet, green, and near infrared radiation of 250 nm, 546 nm, and 1300 nm respectively.

b. Microwaves in the kitchen appliance are tuned to 2450 MHz, an FM radio station broadcasts at 91.4 MHz, and an AM station uses 880 kHz. At what wavelengths do these operate?

8.6 PHOTONS

A useful energy unit for atomic scale processes is the electron volt (eV). In terms of SI units,

$$1 \text{ eV} = 1.6021733 \times 10^{-19} \text{ J}.$$

The reason the eV is useful in optics is that visible light photons have energies in the range of 2 eV to 3 eV. Even more important is that the energy released or required in chemical reactions on the molecular scale is in the same range. In chemistry textbooks the energies involved with exothermic and endothermic reactions are stated on the basis of moles of reactants (joules/mole). This can easily be converted, using Avogadro's number, to energies per molecule. The conversion is

$$1.00000 \text{ kJ/mol} = 1.03643 \times 10^{-2} \text{ eV/ molecule}.$$

Since chemical reactions usually involve hundreds of kJ/mole, this implies several eV/molecule. A single photon with an energy of a few eV can break a chemical bond. X-rays too are photons but with energies of thousands of eV. They can therefore do a lot of damage on the molecular scale. That is why too much ultraviolet from the sun is a cause for skin cancer and why X-rays can kill living cells.

The conversion factor between the energy of a photon E measured in eV to its wavelength λ in a vacuum measured in meters is

$$E = 1.2398455 \times 10^{-6} / \lambda.$$

8.6.1 To get a feel for these magnitudes it is useful to calculate a few conversions between units.

a. The bright red beam from a HeNe laser consists of a stream of photons of wavelength 632.8 nm. Calculate the energy of each photon from such a laser in both eV and in joules.

b. A similar HeNe laser for classroom use has a power output of 2.0 mW. How many photons per second does it emit?

8.6.2 In modern optical fiber communication, the light source is a solid state laser (laser diode) typically operating at 1300 nm and at an output power of 5 mW.

a. Does this laser operate in the visible, infrared, or ultraviolet range of the spectrum?

b. What is the energy per photon from the laser?

c. Signals in this type of communication consist of sending bursts of radiation with a duration of 0.3×10^{-9} s. Calculate the number of photons in each burst.

d. There is a considerable loss in launching the laser light into fibers with a typical diameter of 8000 nm. Once in the fiber, however, the losses are extremely low and the signal is down by a factor of about 300 over a distance of 50 km. Suppose that, of the 5 mJ/s output of the laser, only 1/10 enters the fiber traveling toward the detector. If the pulse duration is 0.3 nanoseconds (ns), then how many photons are in each pulse arriving at the detector? How much energy must the detector be able to measure reliably?

8.6.3 The light flashes used in the fastest fiber optics communication systems have a duration of approximately 0.1 ns.

a. What is the distance between the beginning of the pulse and the end of that pulse as it travels in the fiber?

b. What is the length of the pulse as measured in wavelengths if the source is a 1500 nm laser?

c. What would be the fundamental frequency of the signal being sent if it consisted of a succession of identical pulses separated by 0.1 ns? How does that frequency compare to the frequency of the light being used?

8.6.4 The human ear can be sensitive over a range from 20 Hz to 20 000 Hz. Music spans the same range, and radio has to be able to transmit this spectral range. If radio stations used a technology similar to fiber optics

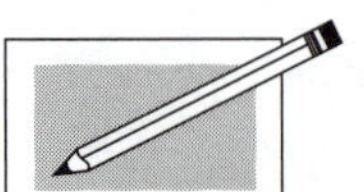

communication, as discussed in Problem 8.6.3, and bursts of radio waves had to have a length of at least 20 times the wavelength you are tuned to, then what restrictions would that place on the music transmitted on the lowest frequency stations of the AM band? Why is FM better from this point of view?

8.6.5 The reaction between hydrogen and oxygen creating water can be explosive. Chemistry texts state that combining one mole of hydrogen gas with 1/2 a mole of oxygen gas into one mole of water releases 240 kJ of energy. How much energy, in joules and eV, will that be per water molecule?

8.6.6 The sun emits radiation over the entire electromagnetic spectrum from X-rays to radio waves. X-rays are completely blocked by the Earth's atmosphere. Most of the visible radiation reaches the Earth's surface and so does some of the ultraviolet radiation. In recent years the intensity of the sun's ultraviolet radiation at the Earth's surface has increased due to the destruction of the ozone layer in the upper atmosphere. Particularly harmful as a trigger of skin cancers is that part of the solar radiation called UV B, defined as ultraviolet radiation in the wavelength range of 280 nm to 320 nm. The part of the solar radiation called UV A, which is ultraviolet in the wavelength range from 320 nm to 400 nm, is less harmful and triggers the body to produce vitamin D and the currently unfashionable suntans.

a. Calculate the range of photon energies for UV B.

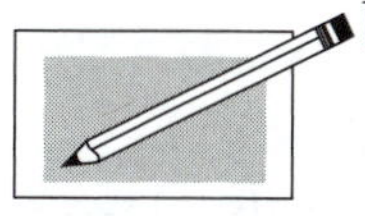

b. The dyes used in cloth and in color photographs are complex organic molecules. Briefly explain why ultraviolet radiation should cause more rapid fading of colors than visible radiation does. Would the same explanation also be valid for the incidence of skin cancer?

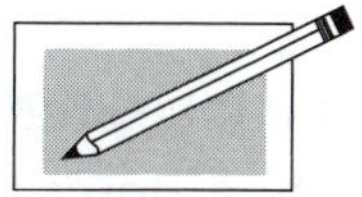

c. On the basis of physical optics explain what the properties of a good sunscreen lotion should be.

8.7 MORE OPTICAL INSTRUMENTS

Chemists, police, and astronomers can determine the chemical elements present in the vapor of a hot object by the characteristic wavelengths emitted. Astronomers look for characteristic wavelengths in starlight. Forensic scientists in police laboratories might take soil samples and burn them in a spark or flame to determine the sample's elemental constituents. The instrument used is a highly sophisticated spectrometer. Two basic types of spectrometers exist for use in the visible part of the spectrum. In one type a prism is used to separate the spectrum into its constituent wavelengths. In the other type a diffraction grating is used. Each has its advantages and disadvantages, which the next two problems explore.

8.7.1 The simplest version of a prism spectrometer is shown in Figure 8.8. The light is assumed to come from a point source. The first lens (objective) changes the diverging beam to a parallel beam and aims it to fall on one face of the prism at normal incidence. Since all incident light is going in the same direction, regardless of where it hits the face of the prism, the direction in which it leaves the prism through another face will be strictly determined by the index of refraction and the angle between the faces of the prism. For a common type of highly dispersive flint glass the index of refraction for violet light (at 404 nm) is 1.84211, for yellow light (at 589 nm) it is 1.78446, and for red light (at 694 nm) it is 1.77231.* (Unfortunately there is no simple formula that accurately relates the index of refraction to the wavelength.) The apex angle of the prism is 29°.

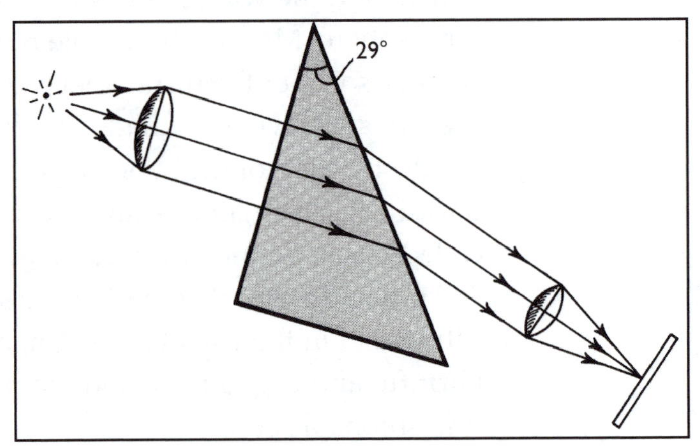

Figure 8.8

a. Calculate the angles through which violet, yellow, and red light are deviated from the direction of the incident light by the prism. Over how many degrees is the visible spectrum spread by the prism? The spread of the spectrum can be increased by a factor of about two when you use a prism with a larger apex angle and when you turn the prism to its angle of minimum deviation.

b. The second lens in this spectrometer, of focal length 25.0 cm, takes the broad parallel beams (each different wavelength at a slightly different angle) and focuses them at some distance from the prism. There the different wavelengths will no longer overlap. If a screen were placed at the focal point of this second lens, then calculate the distance separating the violet and red ends of the spectrum.

c. Calculate what would happen to the spread of the spectrum if a lens of focal length 100 cm replaces the 25.0 cm focal length lens in *part b*. Describe the advantages and disadvantages that will result from the change of lenses.

*Data from the Melles Griot Guide catalog, 1995/96.

8.7.2 Most modern spectrometers used for chemical analysis are based on diffraction gratings. They can disperse the spectrum over a much greater range of angles than prisms can. Diffraction gratings of reasonable quality can now be produced more cheaply than glass prisms, and there is a simple mathematical relationship, $m\,\lambda = d\,\sin\theta$, between the wavelength λ and the angle θ through which that wavelength is deviated. In this equation m is an integer and d is the separation between the slits of the grating. Figure 8.9 shows one way in which a spectrometer using a transparent diffraction grating can be set up. The source is considered to be a point source located at the focal point of the first lens. The light, after passing through the lens, is a parallel beam normally incident on the grating. Each wavelength incident on the diffraction grating is sent off in its own characteristic direction as determined by the grating equation. The second lens gathers the parallel light after it passes through the grating and refocuses it. Since all light of a given wavelength comes out parallel, it is focused at the same point. Different wavelengths come out at different angles and are focused at different points.

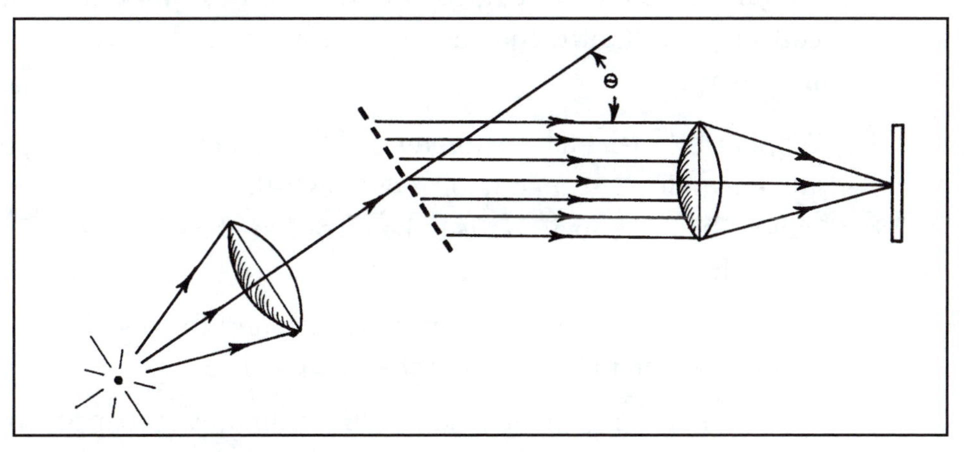

Figure 8.9

For the questions that follow the grating is specified as having 600 lines per mm. The wavelength of light ranges from the violet at 400 nm to the red at 700 nm. The second lens has a focal length of 25 cm.

a. What will be the distance, center to center, between adjacent lines of the diffraction grating? In grating equations this is called the "spacing."

b. Into what angles will the violet and the red wavelengths be diffracted?

c. With the second lens in place and a screen to receive the light, calculate the distance over which the spectrum will be spread.

d. Suppose the light source is a mercury lamp that has strong emissions* at the following wavelengths: 253.7, 365, 404.7, 435.8, 546.1, 577, 579, 615, and 1014 nm. Also suppose that the detector is an electronic device instead of the human eye and therefore is sensitive to ultraviolet, infrared, and the visible part of the spectrum. The detector is mounted at the focal distance of the second lens but its angular position can be changed. Calculate at what angles the detector will pick up light signals.

8.7.3 The combination of a microphone and a large curved metal reflector is advertised to nature lovers as a tool to record bird calls from afar. Not mentioned in the advertisements is that eavesdropping on conversations is also possible with the equipment. Although this particular piece of equipment is designed for sound rather than light, its operation and limitations are best understood through optical principles. The wavelengths of sound are many orders of magnitude longer than the wavelengths of light, but the laws of reflection, energy collection, and diffraction are as valid for sound waves as they are for light waves.

A spherical dish has a diameter of 0.95 m, a radius of curvature of 2.85 m, and is assumed to be a perfect reflector of sound waves. A yellow-bellied sapsucker is to be recorded at a distance of 7.90 m from the dish.

a. Determine the focal length of the dish and the distance from the microphone to the center of the dish.

b. A songbird is a small creature with a small vocal apparatus. Accordingly it must sing at high frequencies compared to human song or speech. Estimate the lowest frequency to be expected from a bird 10 cm tall. (Think of the bird, particularly its throat, as an organ pipe closed at one end.)

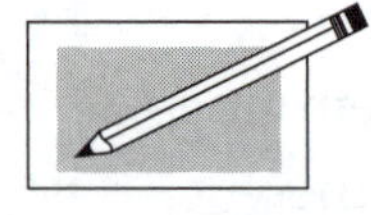

c. Two birds are sitting on the same branch 30 cm apart. One is the yellow-bellied sapsucker to be recorded. The other is a crow trying to sing harmony. The reflecting dish and the microphone are accurately focused on the sapsucker. Over what frequency range will the cawing of the crow be a serious problem for the attempt to record the sapsucker? Explain what is involved.

*Data from the Oriel Optics Corporation catalog.

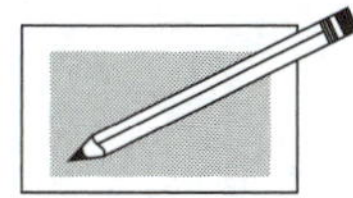

d. A similar apparatus can be used to record human conversations at a distance. Human speech consists of frequencies in the range of 200 Hz to 800 Hz as the fundamentals, with overtones up to the 10 kHz range to lend it distinction. Explain why the dish described above will cause serious distortions in the recorded conversations and why low-frequency background noise will be enhanced compared to high-frequency background noise.

NOTES

NUMERICAL ANSWERS

Chapter One—Mechanics

1.1 Kinematics

1.1.1 a. 1.852 km

b. 0.514 m/s

c. 33.3 km/h; 907 km/h

d. 14.04 m

1.1.2 a. 57.58×10^{11} m; 6.087×10^{-4} ly

b. 82×10^{15} m; 5.6×10^{5} AU

1.1.3 a. 52.6 min

b. 1.8026×10^{12} furlongs/fortnight

1.1.4 a. 31.7 to 1

b. 27.7 km

c. 11.8° N of E-W line, on circle of 27.7 km radius

1.1.5 a. 32.8 m, 11 s

c. 340 m

1.1.6 a. 22.0 min, 27.9 min, 37.9 min

b. 43 km/h, 34 km/h, 25 km/h, all in SE direction

1.1.7 a. 19.44 m/s

b. 189 m

c. 60.6 s

d. 9.93 m/s or 35.7 km/h

e. 8.75 m/s or 31.5 km/h

1.2 Force and Acceleration

1.2.1 approx. 3.2 m

1.2.2 a. 7**g**

b. 115**g**

c. 5.8**g**

1.2.3 a. 100**f**; **f**; **F**; **F**

b. 50**f**; **f**; 0; **F**

c. 39×10^{4} N; 9.8×10^{3} N; 0 N; 20×10^{4} N; 1.5

1.2.4 a. 5.90×10^{-3} m/s^2; 3.43×10^{-5} m/s^2; 3.21×10^{-7} m/s^2

b. 6×10^{-4}; 3.5×10^{-6}; 3.3×10^{-8}

c. 7%

1.2.5 a. 0.628 m/s^2

b. 0.0855 s

c. 2.30 mm

1.3 Circular Motion

1.3.1 a. 113 min.

1.3.3 a. 14.27×10^{11} m

b. 12.77×10^{11} m

c. 1.217×10^{9} m; 5.63×10^{26} kg

1.3.4 a. 1.90×10^{27} kg

b. 10.2×10^{9} km.

1.3.5 a. 29.8 km/s

b. 9.40×10^{11} m; 31.5×10^{6} s; 365 days

c. 5.15×10^{-7} m/s^2

1.3.7 a. 7×10^{-2} s; 0.01 parts in one million

b. 26 610 km

d. 63° N; 63° S

1.4 Work, Energy, and Power

1.4.1 a. 15 m/s

b. 8.3 m/s; 3.1

1.4.2 a. 8.6 m/s

b. 1.7 m/s

c. 2.93 m; 0.54 s

d. 0.68 m; 0.54 s

e. 289 J; 15 W

f. 3.31×10^{3} J; 66 W

1.4.3 a. 46 W; 370 W

b. At 3°: 11 m/s; 14 m/s
At 6°: 16 m/s; 20 m/s

c. At 3°: 180 W; 400 W
At 6°: 330 W; 630 W

1.4.4 a. 2.5 m/s^2; 0.84 m/s^2; 0.42 m/s^2

b. 0.11 s; 0.32 s; 0.63 s
0.63 m; 5.4 m; 21 m

c. 1.7 m/s^2; 0.57 m/s^2; 0.29 m/s^2

d. 500 hp

1.4.5 a. 6.98×10^5 J

b. 72.7 kW

c. 1.15 m/s

d. 113 J

e. 10.9 min

f. 245 skiers; 20 min

1.4.6 a. 7.72×10^4 J; 30.9×10^4 J
19.3×10^6 J; 15.5×10^4 J

b. 13×10^6 J; 52×10^6 J

c. 32×10^6 J; 52×10^6 J

d. 6.1 ℓ/100 km; 9.8 ℓ/100 km

1.5 Momentum

1.5.1 a. 2.79 m/s; 20.2 s

b. 2.1%

c. 0.743 m/s

d. 2.89 m/s

1.5.2 a. 855 s

b. She stops

e. $m = M/4$

1.5.3 a. - 243 m/s, –8.00 m/s

c. Cat: 395 J; 296 J;
- 64.2 kg m/s; – 55.6 kg m/s
Mouse: 1.20×10^3 J; 47.8 J;
10.7 kg m/s; 2.14 kg m/s

1.6 Statics

1.6.1 a. Left: $\mathbf{T} = 135 \times 10^3$ N;
$\mathbf{F}_h = 116 \times 10^3$ N
Right: $\mathbf{T} = 97 \times 10^3$ N;
$\mathbf{F}_h = 91 \times 10^3$ N

b. 30×10^4 N; 67×10^4 N

1.7 Angular Momentum

1.7.1 a. $v_{block} = 2.24$ m/s;
$v_{hoop} = 1.58$ m/s;
$v_{disk} = 1.83$ m/s;
$v_{sphere} = 1.89$ m/s

b. $t_{block} = 1.79$ s;
$t_{hoop} = 2.53$ s;
$t_{disk} = 2.18$ s;
$t_{sphere} = 2.12$ s

1.7.2 a. 11.5×10^3 N m; 1.4×10^{-3} rad/s^2;
3.3×10^{-3} m/s^2

b. 291×10^3 N m; 0.35 rad/s^2;
8.4 m/s^2

c. 0.471 rad/s; 11.3 m/s; 6.5 m/s

1.7.3 f. $v_f = (v_i + R\omega_i)/2$

g. $\omega_i = v_i/R$; $\omega_i = - v_i/R$;
$\omega_i = - 3v_i/R$

1.7.4 b. 5.2 m/s

1.7.5 a. 364 rad/s^2; 22.7 m/s^2

b. 8.40×10^{-4} s

c. 8.00×10^{-3} mm

Chapter Two—Simple Harmonic Motion

2.1 Simple Repetitive Motion

2.1.1 a. 692 kg/s^2

b. 3.65 Hz

c. 2.63 cm; 3.07 Hz

2.1.2 a. 2.02×10^5 N/m; 1.6 Hz

b. 2.02×10^5 N/m

c. 2.25 Hz

2.1.6 a. 335 pF

b. 39.6 pF

c. 4.19 pF to 6.54 pF

2.1.7 a. 8.7731×10^{13} Hz

b. 0.95705 amu; 1.82104 amu;
8.9944×10^{13} Hz

c. 9.6564×10^2 N/m; 9.7152×10^2 N/m

2.1.8 a. 15.872×10^{12} Hz; 16.325×10^{12} Hz

b. 0.536; 0.429; 0.036

2.2 Energy Considerations in Simple Harmonic Motion

2.2.1 b. 8; 15

2.2.2 a. 32.9 J

b. 22.8°

2.2.3 a. 51.4 min later

b. 0.0805 h^{-1}

2.2.4 a. 3.11 s; 39.7 J

b. 9; 0.31 m

c. 0.23 m

2.2.5 a. 9.2×10^4 s^{-1}

b. 0.900; 0.19

c. 9.00×10^{-12} J

d. 10^4

Chapter Three—Fluids

3.1 Buoyancy

3.1.1 a. 29 cm

b. 1.71×10^5 kg

c. 244 m^3

3.1.2 a. 1.29 kg/m^3.

b. 620 m^3; 5.29 m

c. 720 m^3; 5.56 m
666 m^3; 5.42 m

3.1.3 a 1.0%

c. 3.4%

d. 19.8 N

e. 0.92 kg; 2.9 kg; 19.6 kg

3.2 Static Pressure

3.2.1 311 kPa

3.2.2 a. 106.5 kPa

b. 0.53 m

3.2.3 a. –11.1 kPa

b. 33.5 kPa

3.3 Fluid Flow

3.3.1 a. 1.107×10^5 kPa

b. 1076 kg

c. 453 m/s

3.4 Viscous Flow

3.4.1 a. 6.25×10^6

b. 1.96×10^7; 2500 x larger

3.4.2 a. 0.50 ℓ/s

b. 1.6 m/s

c. 1 atm

d. 530 Pa; 2.1 kPa

3.4.3 a. 1.70 m^3/s; 0.92 bbl/d

b. 5.23 km/h; 29 days

c. 2.56×10^6 N; 0.81×10^6 N

d. 1.12×10^{11} J

e. 7.5×10^4 m^3.

f. 1.5×10^3 J; 8.20×10^4 J; 2.2×10^{-3}

g. 1.75 MW; 3.4×10^3 hp

h. 30×10^5 Pa; 7.0 MW

Chapter Four—Heat and Thermodynamics

4.1 Temperature and Energy

4.1.1 a. 0.556 m^3; 0.722 kg; 0.019 kg

b. 0.722 kg; 0.014**g** N

c. 0.817 m^2; 15 s

4.1.2 a. 25×10^3 J/K

b. 12×10^6 J/K

c. 4.7 s; 37 min

4.2 Heat Conductivity

4.2.1 a. 12.44×10^6 J

b. 16.9×10^6 J

c. 7.8×10^6 J

4.2.2 a. 100 W; 54 W/m^2

b. 6.0 W/m^2 C°; 0.17 m^2 C°/W

c. 0.31 m^2 C°/W

d. 37 W; 0.73 m^2 C°/W

e. 0.064 m^2 C°/W

4.2.3 a. 11 minutes; 69 minutes

b. 7 x 10^4 J/s m C°

4.2.5 0.060 atm = 46 mm Hg

4.2.6 121°C

4.3 Ideal Gas Laws and Kinetic Theory

4.3.1 a. 268°C

b. 0.216

c. 8.55 atm

4.3.2

gas(temp)	v_{rms}	v_{sound}
H_2(293)	1906	1302 m/s
H_2(233)	1700	1161 m/s
He(293)	1347	1005 m/s
He(233)	1201	896 m/s
N_2(293)	511	349 m/s
N_2(233)	456	311 m/s

4.4 Heat Engines

4.4.1 a. 0.44; 0.33

b. 1540 MW

c. 0.53; 0.40; 18%

4.4.2 a. 11.2 x 10^3 W; 11.2 x 10^3 W

b. 270 kWh; $22.40

c. 18 kWh; $2.00

d. 18 kWh; $2.00

4.4.3 a. 60.2 x 10^6 MJ; $1.37 x 10^6

b. 54.2 x 10^6 MJ; 66.6 x 10^6 MJ; $1.53 x 10^6

c. $0.55 x 10^6

Chapter Five—Sound and Wave Motion

5.1 Frequency, Wavelength, and Speed of Sound

5.1.1 a. 1.12 km; 2 km

b. 3.45 x 10^4 m/s

5.1.2 a. 248 m/s; 891 km/h

b. 306 m/s to 354 m/s

5.1.3 b. Direct: 32.8 ms, 65.6 ms, 98.4 ms, 131 ms, 164 ms
Indirect: 395 ms, 399 ms, 406 ms, 415 ms, 426 ms

c. Direct: 25.0 x 10^{-6}, 6.25 x 10^{-6}, 2.78 x 10^{-6}, 1.56 x 10^{-6}, 1.00 x 10^{-6}
Indirect: 23.8 x 10^{-16}, 22.8 x 10^{-16}, 21.4 x 10^{-16}, 19.5 x 10^{-16}, 17.5 x 10^{-16}

d. 793 ms; 1.75 x 10^{-30}

5.1.4 b. Direct: 29 ms, 58 ms, 87 ms, 116 ms, 145 ms
Indirect: 210 ms, 227 ms, 246 ms, 267 ms, 289 ms

c. Direct: 30.9 x 10^{-6}, 7.72 x 10^{-6}, 3.43 x 10^{-6}, 1.93 x 10^{-6}, 1.24 x 10^{-6}
Indirect: 11.2 x 10^{-14}, 8.21 x 10^{-14}, 5.95 x 10^{-14}, 4.29 x 10^{-14}, 3.11 x 10^{-14}

d. 450 ms; 5 x 10^{-31}

5.1.5 a. 33.3 x 10^{-6} s

b. 2.7 x 10^{-6} s; 9.3

5.1.6 a. 345 m/s; 33 cm

b. 1046 Hz; 23.9 cm; 250 m/s

c. 1046 Hz; 33 cm, 345 m/s

5.1.7 a. 781.56 nm; 517.23 nm; 221 nm

b. 3.8367 x 10^{14} Hz

5.2 Dissipation and Attenuation of Sound

5.2.1 a. 40 W

b. 200 km

c. 7.4 km

d. 42 W

e. 3303 m

5.2.2 a. 59 dB

b. 62 dB; 60 m

c. 750 m

5.3 Waves of Unspecified Shape

5.3.1 a. 240 km

b. 678 km; 864 km; 672 km; 51°

5.3.2 a. 42 Hz

b. 9.0 m/s; 0 m/s

5.3.3 a. 7.9 h; 4.9 days

b. - 31°C; - 1.7°C

c. 3.0 h; 1.86 days; - 22°C; - 0.01°C

5.4 Sinusoidal Waves

5.4.1 a. $y = -5.5\cos(57t)$

b. $y = -5.5\cos(0.25x)$

c. $y = -5.5\cos(57t - 2.5)$

5.4.2 a. 220 m/s; 14 m/s

b. 63 m; 1.2 m

c. 6.9 m

5.4.3 a. 60°; 30°; 16.2°

b. 53.1 km/h; 29.6 km/h

c. 79.7 km/h, 46.0 km/h; 26.6 km/h, 46 km/h; 8.26 km/h, 28.4 km/h

d. 310 m

e. 49.6°

g. 32°, 14 km/h

5.4.4 a. 1.2 s

b. 12 m; 1.2 s

5.5 Doppler Effect

5.5.1 a. 861 to 900 Hz

b. 880 Hz

c. 0.98 Hz

5.5.2 a. 2.51×10^{-4} W/m^2

b. 4.02×10^{-3} W/m^2; 6.28×10^{-1} W/m^2; 10.0 W/m^2

c. 544 Hz, 484 Hz

5.5.3 $3.18 x 10^7

5.5.4 a. Approaching at 501.2 km/s

b. 0.01%

5.5.5 a. 8.42×10^9 m, 1.68×10^9 m; 213 km/s, 42.7 km/s

b. 434.36 nm; 433.74 nm

5.5.6 a. 404 nm

b. 1.8×10^{18} km

c. 1.9×10^5 ly

d. 2.5×10^{16} km

5.6 Standing Waves

5.6.1 a. 226 N to 3620 N

b. 8280 N; 844 kg

c. 2.96 g/m; 1.66 g/m; 0.932 g/m; 0.587 g/m; 0.330 g/m

d. 138 kg

e. 0.100m; 0.157 m; 0.209 m; 0.254 m

5.6.2 a. 19.6 cm

c. 1230 Hz

5.6.3 a. 3.70×10^{14} Hz; 229 nm

b. 8730

c. 0.093 nm in air

5.7 Beats

5.7.1 a. The A string to 47.0 cm

b. 298.7 or 288.7 Hz

c. 6.09 N

5.7.2 a. 16.96 Hz

b. 5 s or more

Chapter Six—Electricity

6.1 Electrostatics

6.1.1 a. 2.24×10^{-7} N; + or - 4.7×10^{-11} C

b. $E = +$ or $- 1.1 \times 10^6$ V/m
$V = +$ or $- 700$ V

6.1.2 a $Q_{earth} = -5.15 \times 10^{14}$ C

$Q_{sun} = +1.72 \times 10^{20}$ C

b. $Q_{moon} = +1.53 \times 10^{14}$ C or $+3.67 \times 10^{15}$ C

c. $E = -1.14 \times 10^{11}$ V/m
$V = -7.26 \times 10^{17}$ V

d. $E = +3.20 \times 10^{12}$ V/m
$V = +2.22 \times 10^{21}$ V

6.2 Circuits

6.2.1 In order of diminishing brightness: G, C, A and B, D, E, and F

6.2.2 a $5.6 \times 10^2\ \Omega$; $4.7 \times 10^2\ \Omega + 1.0 \times 10^2\ \Omega$

b. 13 Ω; 10 Ω + 2.2 Ω + 1.0 Ω; use series or parallel for 2.2 Ω and higher

6.2.3 a. $100 \times 10^6\ \Omega$

b. 5 at 20 MΩ

c. 7.2 W; 36 W

6.3 Ohm's Law

6.3.1 a. 18.0 Ω

b. 6.67 A

c. 0.607 Ω; 6.45 A

d. 748 W; 25.2 W; 116.0 V

e. 0.382 Ω; 6.53 A; 767 W; 16.3 W; 117.5 V

6.3.2 187.5 W; 182 W; 187 W; 2.75 W; 0.238 W

6.3.3 c. 1.38 kΩ, 10.4 W; 115 Ω, 125 W

6.4 Ohm's Law and Material Properties

6.4.1 a $I = 11$ A, $P = 4.0$ W

b $I = 0.26$ A, $P = 96$ m W; 15 V, 2.4 V

c. $I = 9 \times 10^{-22}$ A, $P = 3 \times 10^{-22}$ W; 4×10^{21} V, 4×10^{10} V. Both are impossibly large.

6.4.2 a. $6.3 \times 10^{-2}\ \Omega$

b. $5.3 \times 10^{-3}\ \Omega$

c. 0.17 m^2; 41 cm

d. 0.17 m^2.

6.4.3 a. 0.080 mm

b. 0.63 mm

c. 250 times

6.5 Non-Ohmic Behavior

6.5.1 a. 4.16 V; 33.9 V; 89.2 V; 3.66 μC; 29.8 μC; 78.5 μC

b. 15.9 s; 71.0 s

c. 1.9 μs; 25 A; 7.5 μA

d. 9.97 s

e. 4.85 μA; 43.7 μW

f. 1.2 W

6.5.2 d. 0.948; 0.494

6.5.3 a. 1.2×10^{-6} A; 1.2×10^{-6} A; 300 A

b. $4.2 \times 10^5\ \Omega$; $8.4 \times 10^5\ \Omega$; $1.6 \times 10^{-3}\ \Omega$; ∞ Ω

6.6 Magnetism

6.6.1 a 2.81 A

b. 1.52 W; 199 m; 1.69 mm; $4.48 \times 10^{-4}\ m^3$

c. 7.80 A; $4.48 \times 10^{-4}\ m^3$

d. 15; 9

e. 14.2 mH; 1.84 mH

6.6.2 a. 8.1×10^6 J

b. 317 H

c. 21.1×10^3 turns; 79.6 km

d. 16.2 Ω; 82.5 kW

e. 108 ℓ/min; 9.4×10^{-2} W vs 82.5 kW

f. 0.60 m

6.6.3 a. 4.52×10^{-3} Wb/m^2, directed along the axis of the solenoid

b. 0.320 N; 0 N

c. 0 N; 13.5×10^{-3} N; 9.55×10^{-3} N; direction perpendicular to long axis of needle and perpendicular to axis of solenoid

6.7 Alternating Currents

6.7.1 a. 27 Ω.

b. 3.5 cm; 1400 m^3; 13×10^6 kg = 13 000 tonnes

c. 930 A; 23 MW

d. 1.1×10^{10} kg = 11×10^6 tonnes

6.7.2 a. 108 V; 0 V; 72.2 V and 51.0 V; 14.4 V and 10.2 V; 6.18 V and 4.37 V

b. 117 Ω

c. 100 W; 22.2 W; 8.89 W; 1.79 W

d. 100 W; 2.29 W; 45.9 mW; 5.64 mW

6.7.3 a. 108 V; 0V; 72.2 V and 51.0 V; 14.4 V and 10.2 V; 6.18 V and 4.37 V

b. $8.8 \times 10^3\ \Omega$

c. 0.295 W; 11.8 mW; 2.17 mW

d. 1.36 μF

e. 1.33 W, 0.295 W; 11.8 mW; 2.17 mW

f. 1314 Ω, 1006 Ω, 938 Ω

g. 38.8 mA, 10.1 mA, 4.66 mA;
34.1 V, 8.89 V, 4.10 V

h. 16.9 V, 1.31 V, 0.27 V;
32.5 mW, 1.95 mW, 83 μW

i. 1.10 W

Chapter Seven—Geometrical Optics

7.1 Rays

7.1.1 a. 5.5 cm by 1.1 cm

b. 9.7 cm by 1.9 cm

c. 1.44 m^2; 7.1 cm

7.1.2 b. 28 cm

c. 5.4 mm

7.1.3 b. 9 times

7.1.4 c. 20 mm^2

7.1.5 a. 8.7 mm^2

b. 2.2 mm^2

c. 0.25 I_o

d. 1/4

7.1.6 a. 5.8 mm^2

b. 1.50 mm by 1.41 mm

c. 5.1 mm^2
3.00 mm by 2.10 mm
1/4

7.1.7 a. 0.527°

b. radius = 1.61 mm; radius = 3.22 mm;
1/4

c. 0.50 mm^2; 0.50 mm^2

d. .061; 0.015

7.1.8 a. 0.533°

b. 8.34 mm^2; 33.4 mm^2; 1/4

c. 0.503 mm^2; 0.503 mm^2

d. .060; 0.015

7.1.10 a. 6.18°; 14.57°; 0.37°

b. 0.074°; 7.4×10^{-3}°; 7.4×10^{-4}°

7.2 Light Levels

7.2.1 a. 0.50 lm

b. 6.0 lm/cm^2

c. 0.018 mm^2

7.2.2 a. 52.5 m^2

b. 0.21 m^2

7.2.3 40

7.2.4 a. 17; 80; 32

b. 154, 15 400 W; 109, 3260 W;
163, 8150 W

7.2.5 a. 23; 10

b. 9.2 kW; 3.6 kW

c. $1600; $630

7.2.6 a. 0.027 lm/W

7.3 Reflection

7.3.1 a. 108°

b. 22.5 x 10^2 mm^2; ellipse

c. 27.8 x 10^2 mm^2; ellipse

7.3.2 a. 2.5 x 10^{-5}

b. 2.5 x 10^{-5}

c. 2.56 cm^2; 2 x

d. 139 cm

e. 34.3 cm

7.3.3 a. 6.2 mm

c. 48.4 cm; 0.72 m^2

7.3.11 a. 47

b. 63 ns; 18.8 m

c. 1.55 mW, 0.45 mW; 40.8 mW

7.3.12 a. 38.0° to 44.98°

7.4 Refraction

7.4.1 b. 24.9°

7.4.2 a. 3.76°

b. 21.9 cm; 36.9 cm

c. 31.4 cm

d. 1.33

7.4.3 a. 90 cm

7.4.4 e. –90° to + 5.9°

7.4.5 e. –5.9° to + 5.9°

7.4.6 e. 0° to 3.9°

7.4.7 f. Twice the angle

g. Up to 50.9° below the horizontal

7.5 Mirrors and Lenses

7.5.1 a. 10.7°

b. 1.03°

c. 10.5 cm

d. 3.4

e. 5.0 cm

f. 1.9

7.5.2 a. Convex

b. 2.20 m

c. 0.855° vs 1.09°

e. 2.15 m vs 2.20 m; 2 x

7.5.3 9.13 cm; + 5.70 cm

7.5.4 a. 0.779°

b. 5.63°

c. 7.22

d. 34 mm; 7.74°

e. 10 x

f. 20 x

7.5.5 a. 0.93 mm

b. 1.11 mm

c. 2.2 mm maximum diameter

d. 0.92 mm

7.5.6 a. 28 mm, 8.0 mm; 55 mm, 27.5 mm; 205 mm, 53.9 mm

b. 3.5 m; 6.9 m; 26 m

c. 52°; 33°; 10°

7.5.7 d. *f*; 10%

e. *D*/2

7.5.8 a. 2 x

b. 1/400 s

c. Either f/8 or 1/5 s

7.5.9 a. 4 x

b. 40 x

c. 20 m

7.5.10 a. 1.785; 1.516

c. 1.51433

d. 58.9 mm; 38.7 mm

e. 22.1 mm; 22.1 mm

f. Flint: 80.18 mm vs 83.65 mm
Crown: 81.44 mm vs 82.84 mm

7.6 Optical Instruments, Multiple Lens Systems

7.6.1 a. 103 mm; 3.4 m

b. 1/1100

c. 1.1 x 10^5 lm/m^2

d. 98.4 mm; 1.2 %

e. 0.13%

f. 2.4%

7.6.2 a. 35.2 cm; 30.4 cm

b. 1:41

c. 29.3 cm

d. 5.0 cm

7.6.3 a. 22.2 cm; – 1.11 cm; – 1/9

b. 205 cm; – 12.5 cm; –1.25

d. 409 cm; 101 cm

7.6.4 e. 129°

Chapter Eight—Physical Optics

8.1 Interference

8.1.1 a. 3.63 cm, 9.35 °

b. multiples of 3.63 cm, not multiples of 9.35 °

8.1.2 91 mm

8.1.3 a. 0.382 mm

b. 0.75 to 1.0 µm

8.1.4 a. 2.627 x 10^{-5} m

b. 44.60

c. 2.24 x 10^{-2} atm

8.2 Diffraction

8.2.1 a. 0.80 mm

c. 3.4 cm stripe, light, and dark fringes

8.2.2 a. 2.2 m; 0.74 m

b. 28.9°

d. 1/2 h

8.2.3 a. 35 cm

b. 7.0 m

8.2.4 a. 8 to 16 mm

8.3 The Speed of Light

8.3.1 a. < 3.7 x 10^{-5} s

b. 3000 x longer

8.3.2 a. 628.7 x 10^6 km; 792.6 x 10^6 km
b. 35.0 min; 44.0 min
c. 9.0 min slow
d. 628.7 x 10^6 km; 772.9 x 10^6 km
35.0 min; 42.9 min
7.9 min slow

8.3.3 a. 0.239 s
b. 0.067 s

8.3.4 a. 4270 s
b. 9 km

8.3.5 a. 1.1 km
b. 5.0 x 10^{-11} s; 5.0 x 10^{-11} s
c. 6.6 m

8.3.6 a. To within 6.7 x 10^{-11} s
b. 30 000

8.4 Polarization

8.4.1 a. 6.2°
b. 0.07°
c. 2.6°
d. 80%, 20%

8.5 Frequency in Electromagnetic Radiation

8.5.1 a. 1.2 x 10^{15} Hz; 5.49 x 10^{14} Hz;
2.3 x 10^{14} Hz
b. 12 cm; 3.38 m; 341 m

8.6 Photons

8.6.1 a. 1.959 eV; 3.139 x 10^{-19} J
b. 6.4 x 10^{15} s^{-1}

8.6.2 a. IR
b. 0.95 eV = 1.5 x 10^{-19} J
c. 1.0 x 10^7
d. 3.3 x 10^3; 5 x 10^{-16} J

8.6.3 a. 2.0 cm
b. 2.3 x 10^4
c. 5.0 x 10^9 Hz vs 2.0 x 10^{14} Hz

8.6.5 4.0 x 10^{-19} J; 2.5 eV

8.6.6 a. 3.87 to 4.43 eV

8.7 More Optical Instruments

8.7.1 a. 63.26°; 59.90°; 59.23°; 4.03°
b. 1.76 cm
c. 7.04 cm

8.7.2 a. 16 667 nm
b. 13.89°; 24.83°
c. 4.78 cm
d. 8.755°, 12.65°, 14.05°, 15.16°, 19.12°, 20.25°, 21.65°, 37.47°
Second order: 17.72°, 25.98°, 29.05°, 31.53°, 40.94°, 43.82°, 47.56°
Third order: 27.16°, 41.07°, 46.75°, 51.67°, 79.54°
Fourth order: 37.50°, 61.16°, 76.21°
Fifth order: 49.55°

8.7.3 a. 1.43 m; 1.74 m
b. 4 kHz
c. Under 4 kHz

NOTES

